REALITY 2.0

A Revolutionary New Look at the Universe and Life

Blue Ocean

WILLIAM TRUAX

ISBN: 978-1-947436-07-7

LCCN (Library of Congress Control Number): 2024922609

For Lynda, Gus, and Rejean

CONTENTS

PREFACE

This is a book about the universe, and it was written for you.

Why, of all things, did I decide to take on the universe? Well, it didn't start out that way.

I'm a curious soul and had heard about the mysteries and paradoxes[1] connected with something called "quantum physics."[2] Maybe you've heard these words too. They didn't mean much to me; maybe something like, "What happens when stuff gets really small, as in subatomic?" Anyway, I got curious and started to look into it.

And, sure enough, there were plenty of mysteries and paradoxes. The more I looked, the less sense it all made. However, it seemed that I wasn't the only one in that boat. The scientists who were dealing with this stuff had some things figured out really well, but in regard to other things, they were completely out to sea. Their situation could be described this way: "We know an Airbus 320 with 138 passengers and 7 crew is going to land at exactly 3:35 p.m. However, we don't know what airport it's going to land at or what airport it will be coming from. We don't even know where it is right now, or even if it exists at all yet. But we know that it will land at exactly 3:35 p.m."

What kind of an answer is that?

---

1. Paradox: any person, thing, or situation exhibiting an apparently contradictory nature. See Notes for citation.

2. According to Caltech, quantum physics is the study of matter and energy at the most fundamental level. It aims to uncover the properties and behaviors of the very building blocks of nature. See Notes for citation.

Scientists know enough about the physics of these super-small particles to enable much of the advanced technology we have today: computers, cell phones, GPS, etc. However, there's no coherent explanation of why or how some of these weird things are happening. They just know they're happening and can make some fairly exact predictions about certain aspects of them, while other aspects are complete mysteries.

This bothered me. I wanted to know more than the fact that a particular airplane was going to somehow materialize somewhere at exactly 3:35 p.m. I wanted to know what was actually going on. I wanted more than an incomplete prediction; I wanted a realistic explanation, if for no other reason than to just make sense of it for myself. I wasn't looking to change the world, just to wrap my head around what was actually happening.

So, I poked and prodded at it, reading more and diving deeper. Volumes of science, math, and philosophy came and went, and the hole only got deeper. Then one day I had a flash of insight. I saw that science had already pointed the way to a good explanation, but that nobody else seemed to have seen it yet. Why? While science may already have given us the answer, we might not have had the ears to hear it.

A word of background. I'm Native American and grew up in America with one foot in each culture. Native American culture tends to be more spiritual, European American more materialistic. Both cultures have their own answers, but both came up short of totally solving the problems of the universe and life. While the Native Americans could be individually very able and highly spiritual, they got conquered by a materialistic culture with better weapons technology and more warriors to wield them. On the other hand, while European American culture has incredible technology that it used to conquer the world and spread prosperity, the people know very little about themselves, about life, or about how to live. (My apologies if I've offended anybody—I'm just callin' it the way I see it.)

Cultures of these types have tended to be mutually exclusive. You were either "scientific" or you were drawn to the spiritual and mystical. Trouble was, if you were "scientific," you tended to reject any reality other than the physical. On the other hand, delving into the spiritual side of things tended to just lead off into the weeds of speculation and opinion. Rather than being locked into one or the other, I can see that both viewpoints have some validity and, more importantly, how they might exist and link together. The disciplines of science might be brought to bear on the world of the intangible, to the benefit of both.

Anyway, I came up with an idea of how to link all these mysteries together and explain what was going on. I put this new model to the test, and it produced results. The more I worked with it, the better it got. It predicted things that came to pass. It answered questions that had previously gone unanswered.

After thinking about this for a while, I came to realize that maybe, just maybe, I had stumbled onto something important. Something you need to know about. Something that would not only help sort out some of the mysteries of science, but in doing so, would provide a new framework to use in deciding how to live, why to live, what to do, and how to do it. After all, science is ultimately the quest for "the answer," for understanding. So, it began to make sense to me that science would provide the jumping off point for a better understanding of not only physical reality but all of life.

Having arrived at this point, I decided to write it up and offer it to you here as an interesting and compelling possibility. It is a rather new and different take on the universe and our relationship to it, but I believe it has the potential to open many doors for science and the humanities. It can be the basis for a new theory of reality as well as new sciences of life.

Depending on your existing education and viewpoints, what I'm saying might seem curious, maybe even off-putting at first; but take a look and see if it doesn't do a better job of explaining reality than the models we've been using for so long.

Personally, I found developing this material to be the adventure of a lifetime (really!). If you come along for the ride, you might experience some of that adventure as well. I promise not to make it too tough. In fact, I've written specifically to make the ideas and concepts we'll be using as straightforward as possible. No big words where smaller ones would do. No unnecessary math or techno-jargon. No self-gratifying displays of literary legerdemain[3] (except maybe that one).

So, could what I'm saying possibly be true? Could science and math lead us toward some sort of deeper understanding of life and reality? I think so, and it's quite a good story.

Read on.

3. A display of skill. See Notes for citation.

INTRODUCTION

IS THERE A HOLE IN THE UNIVERSE?

There's a hole in the universe, or at least in the way we look at it. Or, as a recent PBS video expressed it, "What if reality isn't real?"

As humans, we have subscribed to a number of models of reality throughout history. Each of these models was based on what we knew at the time, plus some creative thinking. We'd have one dominant way of looking at things, which would eventually get old and be replaced by a new way of looking at things, which would itself eventually get old and get replaced, and so on. This has been the heart and soul of human advancement.

Of course, in their day, each of these models of reality was viewed as the absolute truth, and change happened only slowly. As Max Planck[4] once said, "Truth never triumphs—its opponents just die out."

Anyway, what was it that triggered the replacement of these models? What was it that triggered new (and better) theories to come into existence?

In a word, data. Facts. New data would come to light that just didn't fit very well into the old model, and eventually somebody would have to come up with a whole new way of looking at things that encompassed not only all the old facts but all the new ones as well.

What was the telltale sign that let us know when the old model was failing? It started to break down. Given the new facts in hand, it didn't work very

4. German theoretical physicist, 1858–1947: originator of the new physics and Nobel Prize winner, 1918. See Notes for citation.

well anymore. The existing theory of reality turned out to have a hole (or holes) in it, which even a raft of well-crafted explanations couldn't plug.

A great example of this is the flat-Earth theory. The earth sure looks flat from here, and that was good enough for a long time. However, somewhere in the 1400s, mariners and fishermen noticed that no matter how far they went (and some of them went quite a long way), they didn't fall off the edge. They found that the old, well-respected flat-Earth theory had a hole in it—there was no edge to fall off.

This prompted some to think the radical thought that maybe the earth wasn't flat after all, but instead was a huge ball. In a lot of ways, this made no sense, but nonsensical as it might have seemed, it sure did fit the facts a lot better. Columbus, Magellan, and others made this abundantly clear. Simply put, the new theory worked better than the old model, and in doing so, opened the door to all sorts of new possibilities and advancements.

We're at a similar inflection point[5] now. Our existing understanding of this universe has worked fabulously well for us—up to a point. However, there's a hole in our current model of the universe and life. Data coming from the realm of atomic physics simply doesn't work within the old model. If you read enough books about it, you'll see that the way this is expressed is that "quantum physics is full of mysteries and paradoxes." This is another way of saying that experiments done in the twentieth and twenty-first centuries on the smallest and most fundamental scale of matter and energy yielded results that don't make any sense within the current framework.

---

5. A point at which a major or decisive change takes place; critical point. See Notes for citation.

Our social sciences are based on a model of the physical sciences that was outmoded about a century ago, and they're having trouble too. Any subject you look at seems to have at least one major irreconcilable division, with experts on one side shouting opinions, trying to drown out the experts on the other side shouting back at them. Are we angels in progress or barely controlled beasts? Is there enough food to go around, or are we all going to starve? Is the planet heating up? Cooling down? Or is it just doing its normal thing?

How about our lives? Why can't we get control of our lives? Why are we miserable, and why is there no stable body of data to address this? We've tried drugs, mysticism, fifteen different kinds of sex, religion, philosophy, materialism,[6] psychology, government control, and a hundred other paths; but, if anything, we're worse off than we were a hundred years ago. Despite all our efforts, despite all our technology and all our experts and all their opinions, we are not a happy people. Our social sciences have some major holes in them, and our lives are dismal reflections of this unhappy fact.

The purpose of this book is to offer a new model of how the world works and to see if there isn't a way to better resolve some of the mysteries and paradoxes offered up by quantum research. And, in the bargain, maybe we can shed some light on the human condition and how we work as people, and how we might revise our social sciences into a form that provides real solutions to some of the pressing issues of humanity.

After all, if we truly understand how the world works, we should be able to apply that knowledge not only to all the stuff "out there" but also to ourselves. We should be able to resolve not only some of the mysteries of

---

6. A theory that physical matter is the only or fundamental reality and that all being and processes and phenomena can be explained as manifestations or results of matter. See Notes for citation.

how the universe works but some of the mysteries of how life works as well. Maybe we can close the hole in our current model of the universe, bringing more understanding and better life along in the bargain.

CHAPTER 1

A WORD[7] ABOUT WORDS

The beginning of wisdom is to call things by their proper name.
—Confucius

Définissez les termes, vous dis-je,
ou jamais nous ne nous entendrons.

Define your terms, I say to you,
or we shall never understand one another.
—Voltaire

There's a very important issue we need to touch on before we get started. It's words.

This is a big topic; one I'll cover in some detail in the appendix. But here's what you need to know before we get started.

There is one single cause that is responsible for more miscommunication, miscomprehension, miseducation, and the resulting unhappiness than anything else. It's mostly unknown, even though it's been right under our noses all along.

Proper and exact definitions of the words we use are *crucial* to understanding. Sketchy definitions are like getting directions over a bad

7. A short talk or conversation. See Notes for citation.

cell-phone connection, only worse. "Take a left and go *fubbled* yards, then go right twenty *brurpf* (feet, yards, miles?) and you'll be there."

It's not just a matter of missing definitions, but also situations where the speaker and listener have different definitions of the same word in mind. Most English words have multiple definitions. Some of these are a little different from each other, some a lot. So, you can get a totally missing definition, which is like a dropped-out word on a cell-phone call. But you can also get a definition that is just slightly off—enough to corrupt the connection. The word *happiness* is a great example of this. What does it really mean? Short-term elation? A warm feeling of satisfaction? Or a long-term state of awareness and being? Those are three really different things hiding under cover of the same word.

All it takes is one bad definition to render a simple communication totally worthless (or worse). And bad definitions are increasingly prevalent in our society as more and more of us are taught, when encountering a new or confusing word, to "figure it out from the context," or maybe just to ignore it altogether.

Why would I bother to point this out? It's not news that badly defined words impair communication.

What is new is finding out just how severe an effect this has and how hard it hits. It only takes a little imprecision or lack of agreement on the meaning of any one particular word to markedly reduce the understanding of a communication.

As simple as it may seem, this effect can make human advancement difficult or impossible.

All too quickly, the level of understanding of the new incoming communication drops to the level of things the recipient already knows, or worse. When that happens, little or nothing new gets communicated.

Unfortunately, it only takes a little fuzziness to completely bar the communication of important ideas. You can't communicate 100-watt ideas with 50-watt definitions. They simply have to be more precise than that. As misdefining creeps in, the truly significant ideas on which a culture or a technology are based quickly get misunderstood or lost altogether.

You've probably seen cases where some person carefully explained something to another person, only to have the recipient not "get it" or get something *completely* different from what was intended. It's easy to assume this was because the recipient wasn't too bright or wasn't listening or something else. In actuality, it was probably because the speaker and the recipient didn't have enough knowledge, precision, and agreement on what the words in use actually meant. So, the recipient didn't "get" what was being taught. *Clearing up the language and making sure words are well defined can lead to increases in IQ.*

It's not a new idea that lousy definitions impair communication. We were aware of that already. What's new is the idea that *the quality of communication decreases, not as the sum, but as the product*[8] *of the inadequacy of definitions of both speaker and listener.* It's not a nice, linear decrease in understanding—it's where understanding falls off a cliff.

What's new is the recognition of how fast communication becomes ineffective in the presence of poor definitions.

8. In math, the result obtained by multiplying two or more quantities together. See Notes for citation.

For the moment, what we need to know is that words clearly understood and cleanly used are *vital* to thinking, communication, education, and culture. Communication is the single most vital aspect of our existence, and the single most vital aspect of communication is definitions. To that end, this book includes footnotes to provide the proper definition of certain words or explanations of concepts where the meaning might be unclear or unknown.

Some Definitions

Right up front, I'm going to define some terms in the way they will be used in this book:

- The *universe*, as I use it in this book, means the physical universe: "The totality of known or supposed physical objects and phenomena throughout space." There could be (or are) other universes with other rules, but I don't care about them for the purposes of this book.

- *Reality* (for the purposes of this book) means "matter, energy, space, and time." Quantum reality, since it embraces, manifests, or is manifest in, matter, energy, space, and time, is included in this definition of reality.

- STUFF (small caps) means the physical. Anything composed of matter or energy, existing in space and time.

IMPORTANT NOTE

If you feel put off, confused, or vague about something in this book, it may be due to a disagreement between you and me about the definition of a word just prior. Find it and clear it up, and the issue will likely resolve.

CHAPTER 2

A BRIEF HISTORY OF UNDERSTANDING

Any fool can know. The point is to understand.
—Albert Einstein

Wisdom that a wise man attempts to impart always sounds like foolishness to someone else.
—Hermann Hesse

Why do we seek knowledge?

We're looking for understanding.

Since the earliest flickering of awareness, humans have been looking to accomplish certain things:

- Figure out how the world (and later, the universe) works.
- Figure out how life works.
- Figure out how to use this knowledge to resolve the Big Questions of life, such as, what is life? what is reality? why am I here? Those answers are the big prize.

As a race, we have used several different approaches to get this information. All of them have made significant progress. However, each of them has, to date, failed to get us the answers we were looking for. But with recent developments, we have some hope in regard to getting those missing answers.

Religion

The earliest approach to these questions was religion. A deity or someone in an enlightened state attempted to tell us how things really were. New knowledge was brought to humanity. However, this process did not take place without issues.

For example, there were almost always differing interpretations of what the prophet or deity "really meant." After all, the deity or prophet was bringing completely new and different ideas to a population that had almost no experience with those ideas, and they weren't always understood well. The prophet or deity was speaking to a culture that had no words for, and almost no real experience with, these spiritual realities; so, a lot of what was said ended up being stories, parables,[9] or allegories.[10] It would be like talking to the ancient Romans about automobiles and electricity.

Whatever spiritual truths had been brought forth tended to get blended with the not-so-spiritual day-to-day realities of the recipients and were then disseminated in that blended form. Maybe something like "Don't steal (because it makes less of you)" got altered into "Don't steal from our group," which then got altered into "It's OK to steal from *them*." Once these "blends" were disseminated, they tended to be further interpreted (and misunderstood) as they spread.

---

9. A usually short fictitious story that illustrates a moral attitude or a religious principle. See Notes for citation.

10. A poem, play, picture, etc., in which the apparent meaning of the characters and events is used to symbolize a deeper moral or spiritual meaning. See Notes for citation.

Over time, the understanding of the spiritual realities that existed at the core of the religion got weaker, and eventually this weakening caused the religion to fall into disuse. One result is that religion these days is thought by many to be irrelevant or *passé*.[11]

Moral codes were part of this process. Moral codes were an important civilizing factor and were generally introduced into human societies by religion. They were attempts to answer the question, "How should I live?" Without morality, we would not have civilizations today. Things are already "dog eat dog" enough as it is. However, since morality had its origin with religion, and since the religions tended to fall into disuse, moral codes also tended to weaken with time. As these codes faded, they were replaced with all sorts of temporary and subjective (and usually unsuccessful) rules of conduct. Lacking any clear idea of what might be right or wrong and *why* it might be so, these later codes never worked particularly well.

As a result of these and other factors, religion, standing on its own, could be considered, as of the mid-twentieth century, as having failed to get the big prize.

Philosophy

Philosophy's job was (and is) to nail down what we consider to be essential and true. According to Dictionary.com, the meaning of the word *philosophy* is "the rational investigation of the truths and principles of being, knowledge, or conduct." Think about that . . . rational investigation of the truths and principles of being, knowledge, or conduct. That sounds pretty fundamental and important. How could you do or think about anything in life without having nailed down what you

11. No longer fashionable, in wide use, etc.; out-of-date; outmoded. See Notes for citation.

consider to be essential and true? Whatever you consider to be basic truths and principles would affect all your thoughts and actions about life, and getting these things as right as possible would be basic to any progress you hope to make.

Once developed millennia ago, philosophy came to be the senior-most discipline of life. All the other fields of learning—math, grammar, logic, astronomy, aesthetics, and the like—were subordinate to philosophy. Philosophy was an attempt to give all these other activities coherence, purpose, and direction.

Unfortunately, since one idea was apparently just as good as another, philosophy degenerated into something of a "shouting match" between different people and different ideas, all without any sort of central point of reference. Humanity lost interest in this subject, once so essential. Philosophy became a derelict sister to the "real" sciences.

Why? Too much guesswork. Philosophy has historically been presented with incomplete data about life and the world, then called upon to fill in those gaps with educated guesswork. Different people made different guesses, then argued about them. While it was better than nothing, working with incomplete data was always going to be problematic.[12] However, as human knowledge has advanced (mostly thanks to science), more and more of those gaps have been getting filled in—enough for philosophy to once again come to the fore.

Science

This brings us to science. At its core, science is an organized search for knowledge and understanding.

---

12. Of the nature of a problem; doubtful; uncertain; questionable. See Notes for citation.

While the core activity of science is the search for and organization of knowledge, science made the decision early on to rely exclusively on certain types of hard data from the physical universe for its theories. This decision fueled many advances, and science has made huge gains over the last couple of centuries.

At one point, in the late nineteenth and early twentieth centuries, it even looked like science would be able to come up with a comprehensive explanation for everything, including the nature of reality. But later developments brought up all sorts of mysteries and paradoxes that the existing "hard" sciences have been unable to satisfactorily explain, and that have, for the moment, put paid to[13] any sort of workable "theory of everything."

The hard sciences, as they stand, have also failed to deliver the big prize. Why? In my opinion, too much insistence on physical-universe data as being the only acceptable data.

While not uncontroversial, it was shown midway through the twentieth century that this universe is somehow incomplete and that the basic principles governing it could not possibly be proven within it. (Much) more on that later in chapter 16, entitled "Incompleteness." In the meantime, if you insist that the physical universe, the universe of matter, energy, space, and time, is the only possible source of data, the only direct answers you'll ever get are answers about how that physical universe works. What if there's more to the puzzle than that? What if there's something, or some reality, that somehow goes beyond the physical? If that's the case, many of the answers about how life and this universe work will be beyond your reach. And you'll end up with unanswered questions, mysteries, paradoxes, and theories of existence that require miracles to hang together.

---

13. Finish off, end. See Notes for citation.

Building on the Past

Even though religion, philosophy, and science each failed as individual practices to get us the answers we were really looking for, it doesn't mean they were complete failures. Each practice, building on the advances of the others, is what got us to where we are today. And where is that? If we're smart about how we use all three of these disciplines, we're in a position to take a crack at a new (and hopefully better) model of reality.

Science has been good to us. We know a lot about how this universe works. However, it appears that we may have gone about as far as we can go with physical knowledge alone. For example, we can put together wonderful theories of what's happened in the universe over the past billions of years, but what about the universe itself? How did it happen? How did it get here and why? Has it always been here? Much current evidence and theory points toward a singularity[14]—a point where things began. But while they can point at it, they can really say no more. We've reached the limits of physical knowledge. There may be something that goes beyond the strictly physical, which is somehow intimately involved with its operation.

If we want to be able to answer questions about life and this universe, we have to find out about, or know about, life. And the answers to the questions about life aren't necessarily contained entirely within the bounds of the physical universe. You can use data from this universe to get oriented and fill in much of the picture, to learn how to think and to know where and how to look, but some part of the investigation will have to go beyond the purely physical. In other words, we can use the practices

14. A point in which all physical laws are indistinguishable from one another, where space and time are no longer interrelated realities, but merge indistinguishably and cease to have any independent meaning. According to the author, singularity can be thought of as a "zero point" at the beginning (or end) of a universe See Notes for citation.

and disciplines of science to inform[15] philosophy, to give it the data and discipline that's needed, then use that upgraded philosophy to combine what's best and most useful of science and religion to create a new and better model of reality.

We have all the data and all the advances made over the past centuries at our fingertips. We have all the tools of science, philosophy, and religion at our disposal. I think we have enough data, enough experience, and enough expertise to take another crack at answering those questions that have been demanding answers since time immemorial.

Ready?

---

15. To animate or inspire. See Notes for citation.

CHAPTER 3

WHAT IS A SCIENCE?

Sapientia est potentia, *often translated as "wisdom is power."*

In a darkened world no longer illuminated by the light of this center,[16] *technical advances are scarcely more than despairing attempts to make Hell a more agreeable place to live in. This must be particularly emphasized against those who think that by spreading the civilization of science and technology even to the uttermost ends of the earth, they can furnish all the essential preconditions for a golden age. One cannot escape the Devil so easily as that.*

—Werner Heisenberg

Science cannot see where philosophy has not lit the way. And philosophy cannot light anything unless science provides it with fire.

— William Truax

Let your mind drift back . . . It's a bright, sunny afternoon in Greece a couple of thousand years ago. Aristotle is walking around, lecturing his students. The subject at hand is, "Does a large and heavy rock fall faster than a small and light one?" Aristotle responds that of course this is true, because there's more attraction between the large rock and the earth than there is between the small rock and the earth.

---

16. A reference to spiritual reality. See Notes for citation.

Sounds good, but it's not true. There was a time (maybe not so long ago) when the best knowledge of humanity was based on what some authority said. This was fine as long as that authority was thinking rationally based on data acquired through accurate observation. But some of the time, accurate observation wasn't really possible or hadn't yet been done, or perhaps the thinking involved wasn't entirely rational; so, the authorities in question came up with ideas based on what they *thought* should be the case.

This was cured, to a large degree, by modern science. We know that science has made huge differences in our quality of life and is the hook on which we hang our hopes for a better life and better tomorrow. But let's dig a little deeper. What, exactly, is a science?

When we think of "science," we tend to think of various applied technologies, laboratories, experiments, and studies being done, and people in white coats making discoveries and creating wonderful new devices. In that world, "scientific" truths are those that can be proven with evidence from the physical universe. Nothing else measures up.

But the true nature of science isn't so much about doing, making, or even proving things; it's really about understanding things. It's not only about studies and experiments that try to discover new data and relationships; it's about the theories and explanations that we come up with for these new data and relationships. Most of all, it's about the organization of all this data into a meaningful and workable whole.

According to Dictionary.com, the first definition of the term *science* is "a branch of knowledge or study dealing with a body of facts or truths systematically arranged and showing the operation of general laws."

The word comes from the Latin *scientia*, meaning knowledge.

This definition says nothing about technologies, peer-reviewed studies, or cool new devices. Technologies are what you get when you apply a science to a particular area.

In science, the pot of gold that you're after is the most basic principles that govern an area. Once you have those fundamental principles, you can use them to work out or find other less basic (but still very useful) principles. Using these, you can do experiments, get some facts, and work out other principles. You can then use those facts and combine more and more of these principles until you have a body of facts and principles that covers an entire field and can explain everything that happens in that field. For example, once we sorted out the actual laws of motion, we could use them to accurately predict and describe anything that moved. We could shoot a probe to Mars over 130 million miles away and have it land within meters of the target.

While a science spends most of its time acquiring knowledge, it's ultimately all about using that knowledge to find and organize the most fundamental natural laws that exist.

The Pyramid of Knowledge

You can use the idea of a "pyramid" to describe how an organized body of knowledge fits together. The most fundamental or basic data are at the top of the pyramid. When these basic data are put into use, they combine to form a broader and more complex layer of principles and activity that fits just below that first level. These things are then combined to make up even more complex activities, and so on.

As an example of this pyramid structure, all the matter in this universe is made up of three basic types of atomic-scale particles: protons, neutrons, and electrons. That's the pointy end of the pyramid in regard to matter. These three basic building blocks can be combined together in different ways to form atoms of all of the different elements, such as gold, iron, and oxygen. These elements are the next level down. The elements can then be combined to form compounds such as water, sand, or rust. That's one more level down. Even more complex structures can be formed from these things, such as minerals, air, and soil. Going down several more levels, we get to some of the complex chemicals and structures used by living organisms. And things keep getting increasingly complex until it accounts for everything that you could possibly make out of matter. Those three basic building blocks cover the entire huge and complex field and can explain everything in that field.

The same basic pattern holds true for any body of knowledge you care to look at. The hard sciences have generally done a good job on their basics, which is why they work so well. Newton's[17] three laws of motion are some of the basic data of traditional physics, and everything we know or can see about motion (on a nonatomic scale) can be worked out from those laws. All bodies of knowledge in this universe are organized on that same basic pattern. Basic principles or manifestations combine into increasingly complex forms until we have a fully realized universe.

When you're working with a knowledge pyramid like this, there are a couple of ways of applying logic. One is that you can take some known data or principles and use them to compute or predict less-basic data or phenomena. For example, you can use the laws of motion to predict how a propelled object will move. That's called deductive logic. The other thing you can do is look at some data or phenomena, spot a pattern, and figure out what has to be the basic operating principle(s) behind it. That's called inductive logic. In terms of the pyramid of knowledge, deductive logic moves down the pyramid toward the bottom, and inductive logic moves up toward the point. They're both valid ways of figuring things

17. Sir Isaac Newton, 1642–1727, English philosopher and mathematician: formulator of the law of gravitation. See Notes for citation.

out. Deductive logic tends to produce technology, while inductive logic tends to produce basic principles.

The Importance of Principles

The key to any field of knowledge is to identify the most basic principles that are true for that field. If you start with principles that are "sorta" basic, or close to (but not quite) basic, for that field, the solutions that use those principles are eventually going to be partially incorrect. And if you don't have the *most* basic principles, your development in that field is going to end up with some loose ends—things that you can't explain without getting really complicated, or for which your explanations are incomplete, or that show up as paradoxes or mysteries.

As you get principles that are increasingly basic, they will explain more things and explain them more simply, and the most basic principles will be able to be used to explain and/or predict all the phenomena of an area. Each field of human knowledge has a set of basic principles that are senior to that field, and it's those principles that define what's true or what works for that field. They are the rules by which that game is played—the soil from which the plants of the field grow. While the rules define the game, they are never part of the game itself. These fundamental and unchanging rules are called axioms.

What Qualifies as Knowledge?

So, how do we get knowledge and what qualifies as knowledge? To do science at all, you have to acquire knowledge or data. In order to do this, you have to know how to tell what data is valid. The way science has worked, "scientific" truths are those that can be proven with evidence from the physical universe. If you can't do an experiment, do it anywhere, do it with anyone, and get the same results clearly demonstrated in the physical universe each time, it's not a "scientific" datum. This works great for the "hard" sciences but has been less effective for the human sciences.

The hard sciences work a lot with quantities. The human sciences work mostly with qualities. If you have the basic principles (the axioms) of either, you can use logic and other tools to help sort things out from there.

If you insist that the only truths that can be known are those that are proven by physical phenomena, you're getting yourself into a box where the only things you allow to be "true" are phenomena that have physical causes. That's great for a lot of things, but it leaves some really important questions totally unanswerable. For example, scientists can come up with an explanation for almost all of what's happened since a postulated Big Bang occurred, using nothing but physical data. But what about the Big Bang itself? Where did that come from? And what about the human sciences? Are all of the causes there strictly physical?

If we want to be able to work in the human sciences, we have to be willing to look beyond the bounds of strictly physical causes. There's more to life than just STUFF.[18] Moreover, we have to do this while keeping our methods straight and our observations clear. If we can do that, the hard sciences can show us the way to the human sciences, allowing us to work with knowledge that encompasses more than strictly physical data.

A true science is not just a collection of data, a pile of exhaustive (but somewhat unrelated) studies, a set of procedures, brain-busting math, or a bunch of interesting technology. It starts with the most basic principles (axioms) of that field, organized and expressed so that other principles can be predicted or derived from them, and so all the phenomena of that area can be explained or understood and handled in a straightforward manner.

---

18. STUFF. The physical stuff: Anything composed of matter or energy, existing in space and time.—William Truax.

A true science is the key to real understanding, and it can embrace the nonphysical (thoughts, ideas, etc.) as well as physical phenomena.

So, we can acquire knowledge, and once we have it, we should probably work with that knowledge using logic and reason. What's so special about logic and reason?

CHAPTER 4

WHY REASON?

Science uses reason and logic, which is great for dealing with STUFF. But aren't there other, and maybe better, ways of thinking about people and things related to people? Shouldn't feelings be part of the equation?

The Romantic movement of the nineteenth century created magnificent works of music, literature, philosophy, and art that were among the finest ever created. The principle that drove them was the idea that the feelings, the perceptions, and the emotions of the individual were the most direct route to truth. They didn't have much use for dry and "lifeless" reason, feeling that they were on a more direct route to the center of all things.

As plausible and vital as that movement looked, it turned out to be a dead end. They eventually came to believe that truth was beyond the reach of mere mortals. The best we could ever do as a race was aspire to an unreachable state of enlightenment. The problem with that was, though true enlightenment was unreachable, it was only by reaching it that we could ever find out how our world works. Ugh—no irony there. Fortunately, it turns out that if you really want to find out how the world works, you can get somewhere if you use reason.

Why reason?

Because it works.

Reason has been a valuable tool for humanity. It has enabled whatever mastery we've achieved over the physical universe. It's brought us science and technology. It's enabled us to feed and clothe billions of people, and has brought us prosperity as well as victories over death and disease.

On the other hand, it hasn't yet given us the whole answer. Our sciences are incomplete. We're polluting the planet with no end in sight. We haven't yet figured out how to build economic systems with sustainable prosperity for the many. We don't know how to live truly healthy lives, or even how to attain happiness.

Whenever an old solution has come up short, there's been a temptation to toss it all out and start fresh. After all, it failed, right? Along those lines, modern society has a number of voices calling for the abandonment of reason. They feel we would be better off with other ways of dealing with things. From where they sit, they wonder why one particular thought process or concept of reality should necessarily have to apply to all of us. Why can't we all have our own truths?

In a broad metaphysical sense, I suppose we all might have our own truths. However, the most pressing problems of humanity are not problems that require lots of metaphysics to solve. The most pressing issues of humanity are those issues that are here and now, in this universe.

And this universe runs on, and is built on, logic. That's the entire explanation of why reason works in this universe.

Why is it built this way? That's one I can't answer. What I do know is that the physical universe can be viewed as a system of interrelated parts. These parts interact according to certain rules. One such set of rules is arithmetic. Arithmetic is the set of rules that governs quantities of things. In this universe, 1+1=2.

There are lots of other sets of rules as well. They're frequently referred to as laws, such as the laws of motion, the laws of thermodynamics,[19] etc.

All of these sets of rules are related to a single more basic set of rules. This basic set of rules governs how all the other rules work and interrelate. These more basic rules are the rules of logic.

Logic is nothing more than an expression of how this universe works at the most fundamental level. It's how anything that is part of this universe is put together and operates. As an example of a rule of logic, "What a thing is, it is." Another example is, "Either something is, or it isn't." There are other rules, but you get the idea. They're pretty basic.

If we want to make things work in this universe, we have to acknowledge, know, and use the rules of logic. When dealing with the physical, logic is what works. That's because it's how the STUFF is made.

Reason is the practice of gathering useful data and then using logic to figure things out. It works because it's using the same rules that govern this universe, and that's the only reason it works. Wherever reason has fallen short, it's because it was operating off incomplete or erroneous data or was not being applied correctly.

Reason is not entirely instinctual. It's a skill that has to be practiced and learned. Not only do you have to learn how to do it correctly, but you also have to realize that you're going to be dealing with all sorts of potentially false data, prejudices, preconceived notions, emotions, and such. You have to work out how to ensure the data you're using is true. You need to understand that what you want or are willing to see can significantly influence what you actually see. You have to be willing to address the fact

19. The science concerned with the relations between heat and mechanical energy. See Notes for citation.

that you or others may be heavily invested in seeing that a certain thought process comes to a particular conclusion. That's why science came to insist on using physical data only. It's hard to call something white when it's clearly black, no matter how much you wish it were white. It was a way to control false data. The introduction of false data is one of the ways reason can get sidetracked and come up with answers that are almost (but not quite) right.

The trouble with "almost" answers is that they eventually yield bad results. That has given reason something of a bad name.

Much of the progress of humanity over the past centuries has been based on getting more data and better data. Reasoning can only be as good as the data it's based on. History is loaded with examples of incomplete or incorrect data. Logic got applied to this bogus data, and some of the explanations and systems we came up with were, well . . . less than ideal.

The flaw has not so much been with our reasoning as it has been with our data. However, we have been increasingly getting our facts straighter and figuring out how to better use those facts. It looks like we finally and actually have all the pieces necessary to assemble some important parts of the puzzle of life.

In summary, reason and logic are tools we need to use to resolve the questions of life in this universe. Reason has failed in the past due to bad data, but thanks to advances in science and math, such as quantum physics, we may now have enough data to take on this universe and our place in it.

Talking about reason and data and how we're ready to take the next big step is all fine and well, but let's look before we leap. How will we know if our data is good? What is our standard of proof?

CHAPTER 5

WHAT IS PROOF?

Ever play three-card monte? It's a street gambling hustle. It looks easy. You're shown three cards, two black, one red. These cards are then turned over and shuffled around face down on a small table. Your challenge is to keep your eye on the red card. If you can do that, you win $20. But even though you flawlessly track that red card, the card you track always ends up being black.

How does that happen? Simple. It's a beginner's sleight-of-hand move to replace the red card with a black one at the beginning of the shuffle, then have you diligently track the wrong card through all the dazzling moves. And bang—you're out $20.

Proof can be like that. You isolate what seems to be the right factor, you track it flawlessly, and you end up in the wrong place. As a race, we have made this mistake over and over. Yet we have learned. The way we do science and math these days is an outgrowth of being fooled in the past, imposing discipline so we won't get fooled again.

Even with modern discipline and procedures, we still have issues with nailing down what's true, especially in the social sciences. One study says one thing, while a different study says the opposite, and then the talking heads argue about it, asserting that their truth is the real truth (and the others are incompetent and/or liars). How do we know whom to believe? How can we know what's true? What is proof?

How do you prove something? Proof is important, because without proof, it's hard to be sure about knowledge. Is something true, or does it just look like it's true?

Depending on who's doing the talking, there are different standards of proof. Mathematicians use nothing but self-evident axioms and pure reason to prove things. Hard scientists will only accept repeatable observations of physical-universe data. Social scientists use studies and statistics, then throw in some interpretation. Math gives the highest level of certainty, followed by the hard sciences, then the social sciences.

Math, Physical Science, and Social Science

For the most part, mathematics doesn't need observational data but works with pure reason. Given some basic principles that everybody agrees on, the mathematicians can prove things from there on out with logic alone. They don't need to resort to other observational data that could be mistaken or misinterpreted.

The physical (hard) sciences work with STUFF. They prove things by doing experiments on STUFF. They isolate one thing, one variable, then test it out and see what happens. If the test results can be reliably duplicated by others working independently, it's considered to be true. They then draw conclusions using those test results.

The "soft" (or human) sciences work with data related to people or other living things. Trouble is, it's hard to isolate just one variable when you're working with complex living entities. These scientists have to deal with lots of variables and might not even know what some of them are. This is why the soft sciences generally work with studies instead of single experiments—so that they can use statistics and interpretation to try to isolate the effects of the one variable they're testing for.

The level of certainty of the social sciences is not as good as the hard sciences and leaves a lot of room for opinion and interpretation. When this happens, we end up with bodies of knowledge or opinion that don't produce results as well as they should or can't handle the problems they should be able to address. They're not as workable as they should be.

This has led the hard sciences to look down on anything that can't be proven with indisputable data from the physical universe, and math to look down on both.

"How do you know?" The mathematicians seem to have the best shot at certainty. The hard scientists come next because they're working with straightforward physical facts. The processes of the soft sciences leave room for slanted observation, low-quality data, and biased interpretation of results.

Proof?

Proof is an interesting subject. One of the more skilled mathematicians of the twentieth century was Kurt Gödel. He was Einstein's best buddy at Princeton and was able to use mathematics and reason to prove some world-changing things. One of his biggest accomplishments was his "incompleteness theorem." Basically, this said that any system[20] that uses logic and quantities of things can never be complete, in and of itself (more about this in chapter 16). This had implications far beyond mathematics. Gödel's theorem applied to everything in this universe that was subject to

20. A system, as used here, means an assemblage of things or parts that form a unified whole. This could be physical parts, or an assembly of facts, principles, and doctrines in a particular field of thought. Arithmetic is a system, as is any form of mathematics. Also, systems are any science, any form of government, any company, any culture, even this universe.—William Truax.

the laws of logic. Incompleteness is true for math, but it's equally true for fields like science, language, government, culture, and philosophy.

Any field of knowledge or action (system) that uses quantities and is based on the principles of logic has to have certain basic principles that are true but that can't be proven (in a mathematical sense) by any data within that system, even though the data is interpreted based on these principles. Those basic principles establish that field and establish what can be true in that field. These basic principles are usually described as being "self-evident" and are called axioms.

Hmmmm . . . we need basic principles in order to reason properly and establish a science, but we can never mathematically prove these principles. We just have to accept or agree to them.

This creates an interesting dilemma. If the axioms of a field are mathematically unprovable, how do we know we have the right axioms? We can't prove (or disprove) them. How do we know we didn't track the wrong card? Here's how: We make observations of the physical, then use reason and logic to figure out the best explanation for these observations. We then use that explanation to create some trial axioms for that field, apply them, and see if they work. If they work, we see if we can refine them. If they don't, we try again with some new axioms. Francis Bacon[21] wrote this up centuries ago. The proof is in the pudding.

Summary

Math, hard sciences, and the social sciences have different levels of working proof, with the proofs of math being the most reliable, and the hard sciences being the next most reliable.

21. English philosopher and statesman, 1561–1626. See Notes for citation.

Advanced math shows us that the axioms of this universe could be said to exist "outside" of it. It also says we can't use data from within this universe to mathematically prove anything outside it. Its axioms are not subject to absolute proof with the data available to us. That being the case, anything derived from these axioms is ultimately also going to fall short of *absolute* proof. As a result, there is no such thing as an absolute proof of a proposition[22] in this universe.

So, how do we get trustworthy axioms for this universe? If there's no way to obtain absolute proof, how do we get axioms for this universe that work uniformly and produce dependable and desirable results? Fortunately, people have been thinking about questions like this for a long time, and we have some good rules of logic and reason that can lead us to workable answers, so long as our data is accurate. We have rules that allow us to tackle questions like this one, and the purpose of those rules is to allow us to find the best possible answer. The key is workability.

We can use data from the hard sciences and combine them with reason to formulate axioms that can then be compared to life in this universe to see if they work and explain what's going on. If we approach the problem carefully, openly, and intelligently, it's possible to use the tools of math, science, and philosophy to crack the code. We can beat the game and find the red card.

---

22. A thing, matter, or person considered as something to be dealt with or encountered. See Notes for citation.

CHAPTER 6

A WINDOW ONTO TRUTH

Knowledge of the lowest kind is ununified knowledge; science is partially unified knowledge; philosophy is completely unified knowledge.
—Herbert Spencer

Humanity has been looking for truth since, well . . . humanity. There have been some notable successes, but things usually haven't worked out quite as well as they might. Recently, we've begun to wonder whether it's even possible to open a window onto truth.

Yes. It turns out that there's one activity behind almost all the intellectual achievements of humanity, one which we can put to good use in our search. However, most modern people don't know what it is.

We can use this tool to think, to clarify, to evaluate data, and, ultimately, to find truths. It's the most important tool we have. It's philosophy.

For many centuries, philosophy was regarded as the senior discipline of life. It was the pursuit of Truth itself, the discovery and use of those principles that would bring about harmony in all actions and meaning in life. What was true? Why should things be done? What was good? How do we find and know truth? All the big questions of life were questions of philosophy, and the answers humankind came up with for those questions determined how they lived and what they did. Their answers determined the bones and sinews of their culture.

Many attempts have been made to answer "the big questions." Many philosophers have stepped up, taken the data of their day, and put forth the truth as they understood it. However, as fast as one thinker could advance a plausible idea, another emerged to shoot it down. One person came up with the bright idea "X" (for example, all humans are fundamentally perfect), and another came up with "Y" (for example, humans are irredeemably imperfect), and they formed competing "isms." Trouble was, both were improvisations based on incomplete or bad data, and so both ended up being somewhat half-baked and, in the final analysis, unworkable. This was how philosophy degenerated into something of a "shouting match" between different people and different ideas, all without any sort of central point of reference. Since one idea was apparently just as good as another, things got crazy and stupid. Humanity lost interest in this subject, once so essential.

Philosophy became a derelict sister to the "real" sciences and to the modern social sciences. Ask a person in the street about philosophy today, and they'll either stare at you blankly, tell you it's all a pile of meaningless speculation, or explain about their own unique "philosophy of life." Hardly anyone seems to regard philosophy as a serious endeavor that has any sort of shot at truth. Not only have the big questions of life gone unanswered; we've also gotten to the point where we think it's unrealistic or naive to ask these questions at all. We've pronounced philosophy dead as a meaningful undertaking.

While philosophy failed, the "real" sciences were able to produce repeatable results. They relied almost entirely upon a form of materialism as their basic premise. Since they were so successful, we gave up the search for that elusive, intangible illusion called Truth, and instead embraced the good solid realities of science and the material world.

However, there were advances and radical new ideas that popped up on humanity's radar after philosophy took a back seat to materialism. Some of the most notable of these are contained within atomic physics. The developments of physics in the early and mid-twentieth century undermined the basic viewpoints of materialism, showing that it was just another "old," outmoded, and limited way of looking at things. Astounding as it may seem, physical science seems to have shown that living beings or consciousness is somehow involved with reality, somehow involved in the nature and maybe even the creation of the seemingly solid world around us. However, theoretical physics tends to exist in its own little universe, and much of the knowledge they gained has been something of a secret until recently.

Out here in the world, separate from the atomic physicists, we grew ever more dependent on the materialist view of existence. However, materialism treats life as an afterthought and lacks any sort of organizing principles based on the purposes of life. Since materialism allows us no purposes as living beings, it's not surprising that we have become miserable as a society. We have come to "know" that there's no "real" meaning to life. With no grounding in any concepts but those of the material world, we believe life itself to be nothing more than a brief and inconsequential flickering of physical phenomena. We therefore find ourselves increasingly operating according to physical laws, including the inevitable progress of disorder and the eventual "death" of us and the universe. That's what philosophers call a "dead end."

Given all this, it's really no surprise that our leaders, our best minds, and our best technologies all seem to be slowly converging on our destruction or degradation. We're subject to an ongoing series of disasters, which we are increasingly unable to control or address. But it's just the way of the world, right?

This is not the way of the world, nor the way the world needs to be. These are symptoms of cultures that, lacking meaning and a coherent viewpoint on life, have lost their way. These are symptoms of cultures that have outgrown their organizing principles and are now just "waitin' to die." Without some stable central agreements, they can do nothing more than stagger down the road toward a doomed tomorrow. They need a review and renovation. Much of what they have is good and usable, but these cultures and their fields of knowledge need a serious update.

Humanity has a rich intellectual history, as well as a rich history of what happens to peoples and cultures after they adopt certain ideas. There are many interesting pieces of the "truth puzzle" that have come up over the years, although some have been forgotten, invalidated, or overlooked. You can look at history as a laboratory for those ideas. You can find a time when some culture adopted idea "X" and see what happened. (That, by the way, is one of the reasons it's important to know history. It's not merely a lifeless recording of dead events. It's the laboratory of human cultures.)

In some ways, our current scene looks very much like that of sixteenth-century Europe when natural philosophers[23] began to undo the old belief systems and lay the foundations for modern science and society. In doing so, they had to take on the established truth of their day. They had to take on the Church and its doctrine.

Our culture has an accepted doctrine too. But our doctrine is not defended by a domineering church. Our prevailing doctrine is a pervasive and intolerant materialism, defended by its own high priests who proclaim that "there is no reality but material reality," and anyone who dares to think or speak differently is ridiculed as being "unscientific" or simply shouted down.

---

23. Natural philosophy: (now only used in Scottish universities) physical science. See Notes for citation.

Even so, truth is emerging. Many pieces of the puzzle have been put into position by generations of thinking men and women, some long ago and some just recently. Fortunately for us, philosophy did not cease to exist or slide into total uselessness and has now commenced a renaissance. Philosophy is the only activity that can step back, take in the big picture, and finally solve the puzzle. While our current sciences can tell us all sorts of specific data about trees, philosophy can tell us about the forest—what it is, why it's there, and how it got there.

We can pick up all the puzzle pieces, look at them newly, and arrange them into a picture that brings us closer to truth. Order can be brought to the chaos of competing ideas and schools of thought. Outfitted with workable data, philosophy can now take its place as the senior pursuit of humanity. It can use the hard data from math and science, plus the tools of logic and reason, and, in doing so, can provide us with workable axioms to arrange our sciences.

And that is the place of philosophy—giving shape, direction, and integration to all the sciences, all the math, all the knowledge, of humanity. "Philosophy is completely unified knowledge."

OK. Having established that, here's a question for philosophy: If nothing can be proven absolutely, is there anything we can know with certainty? Yup, for sure. There is one thing. And it's the first piece of the puzzle.

CHAPTER 7

WHO'S THERE?

Ever see one of those mystery movies, or maybe read a story, where the truth is slowly revealed in bits and pieces, but just when you start to think you've got it figured out, the writer switches it up on you and your assumptions turn out to be wrong? Like in *Star Wars* where the villain, Darth Vader, turns out to be the father of the hero, Skywalker?

Life is like that.

This has been a problem for us as a race since we first started down the road of figuring it all out. Every time we think we finally have it, the rug gets pulled out and it's back to square one. It's enough to make you think, "Maybe there is no truth." Our mission here is to use philosophy as a tool to develop workable truths. However, one of the problems with philosophy is that within the discipline, it has seemed like any idea is just as good as another. Is there any idea that somebody can't come along and deny? Instead of committing to another set of false starts, can we find one thing, anything, that we know is true? Is there anything that we know is true beyond a doubt? Something we can base our thinking on?

It turns out there is.

One philosopher, scientist, and mathematician, a few centuries ago, studied and read and came to a similar sort of confusion, then decided to accept none of it. He decided to doubt everything, then went on a search for any datum that would be undeniable.

His name was René Descartes. It wasn't until Descartes did his work that anyone had any sort of an indisputable datum with which to approach life. He succeeded in isolating one datum that could not be disproven, pushed aside, or invalidated.

That one, single undeniable datum was "*cogito ergo sum*," or "I think, therefore I am." Prior to that time, anyone who had any sort of idea about life or reality could be shot down or picked apart by some opposing -ism; but here was an idea that nobody could shoot down.

When you look at it for a moment, it's obvious. Since I'm thinking, there must be something or somebody doing that thinking. We don't have to deal with the broad metaphysical implications of how that thought is occurring or the ultimate validity of those thoughts in order to achieve the goal of a better life and culture. Those questions belong to other areas of pursuit. The fact remains that something exists and is doing some thinking.

Descartes dealt with this by establishing two very different categories of existence. The first was "that which thinks" (*res cogitans*, or "thing thinking" in the original Latin), and the second was "that which exists in (three) dimension(s)" (*res extensa*, or "thing extended"). That which thinks was implicitly nonphysical—its existence was not necessarily contained within three-dimensional reality.

He tried to come up with an explanation of how "that which thinks" could affect "that which has dimension" but, lacking data, failed. Because there seemed to be no way to show how a nonphysical entity (which he characterized as "the mind") could influence or control a physical entity (the body), this created a problem that has been called "the mind-body problem." Scientists and philosophers eventually dropped his formulation altogether and have spent the last 150 years or so trying to show that the mind is actually physical, without much success.

That brings to light some interesting questions about what's doing the thinking. Is it pure mind? Is it some other nonphysical thing that has and uses a mind? Is it some sort of thinking or computing matter? The answer that is most accepted these days is that the material brain is the seat of thinking and consciousness, but no one has ever been able to work out just how the thinking occurs.

Even though the idea is currently widely accepted, why is it so much easier or better to consider that matter is capable of thinking than it is to consider that something nonmaterial is doing the thinking? Matter that thinks is a pretty incredible concept, and one with lots of issues.

My experience (and Newton's law) is that matter is pretty inert. It doesn't go unless something shoves it. If so, what's shoving it? And, if we accept that "some" matter thinks (for example, our brains), why is most matter nonthinking? And how is it that our brains do this thinking anyway? Nobody's been able to show how that gets done. And, if we accept the concept of thinking matter, where (and how) do we draw the line between matter that thinks and matter that does not think? If that's true, what causes some matter to think but not other matter? Or, does all matter think?

Anyway, there are all sorts of unresolved issues with the idea of thinking matter. So, why is it that many people believe that thinking matter is "obviously" acceptable, while agreeing that the existence of something nonmaterial doing that thinking is so "obviously" unacceptable?

Let's have a closer look at this. What's the definition of thinking? With a nod to Dictionary.com, it could be defined as "the conscious employment of a mind in evaluating or dealing with a given situation." Ideally, this would be done rationally and objectively. However, there's no law that says minds have to always be rational. And what does *conscious* mean? That definition is straight from Dictionary.com: "awareness of one's own existence, sensations, thoughts, surroundings, etc." So, thinking is the employment of a mind by a consciousness in order to evaluate or deal with some sort of situation (such as living, a threat, whatever). It could (and should) be rational, but it doesn't have to be.

In terms of sorting out what, exactly, is doing all this thinking, we're left with a stark choice. That which thinks is either material or nonmaterial. Either we postulate that matter has somehow, by chance and by itself, worked out rational thought, or we postulate that the source of thought is nonmaterial. Personally, I don't see that postulating matter that thinks (and moreover, can think rationally) is any more likely than postulating a nonmaterial potential that does the thinking.

In 1927, J. B. S. Haldane (founder of a branch of evolutionary theory and an avowed materialist) wrote: "It seems to me immensely unlikely that mind is a mere by-product of matter. For if my mental processes are determined wholly by the motions of atoms in my brain I have no reason to suppose that my beliefs are true. They may be sound chemically, but that does not make them sound logically. And hence I have no reason for supposing my brain to be composed of atoms." Interesting point, that. If my mind is purely material, there's no way to show it's rational. And if I believe it's rational, there's no way to show it's purely material.

Neuroscientists have had their issues with trying to substantiate thinking matter. Every single attempt over the last few centuries to show that matter does think, or is even capable of thought, has failed. While they can show that the nervous system and brain are used in thinking, what they can't show is that the brain is what is doing the thinking. You could look at it as being similar to computers and thinking. Is the computer an aid to thought processes, or is it the origin of those processes?

"Speaking of computers, what about AI?" you might ask. There's a machine that's doing some thinking, right? Might it not somehow develop consciousness? Well, not really. It's just processing the algorithmic instructions that a thinking being created for it. It can get really complex and subtle; it can be made to mimic a living entity, but it still requires a creator and a program. And if you assert that humans are just biological computers processing their own sort of algorithms, that begs the question of who wrote their program? Whose algorithms are these?

That brings up the question of creation. When we look at the activities of humanity, what we are goes beyond mere thought. We also create. How does one explain creation? Somebody had to create this book and the ideas in it. Somebody had to create the music you hear, the art you see, and the very idea of the cities we live in. Not only are we talking about thinking, we're also talking about creating. So, the question is not only, "Can matter think?" but it's also, "Can matter create?"

How is it that matter that thinks and creates is more plausible than something nonmaterial doing these same things? In short, it's not. They're both "outrageous" when viewed in the light of current science. While there may be no recognized proof of a thinking nonmaterial potential, there's also no proof whatsoever that matter can think, much less create.

Hold on, you might say. Isn't it settled science that life came into being in some primordial soup, then evolved into its current form? In a word, no. This theory is *very* unrealistic. According to Fred Hoyle (British astronomer and founder of the idea that heavy elements are created within stars), "Life cannot have a random beginning . . . the trouble is that there are about 2000 enzymes,[24] and the chance of obtaining them all in a random trial is only one part in $10^{40,000}$, an outrageously small probability that could not be faced even if the whole universe consisted of organic soup." By way of comparison, there are about 10^{80} subatomic particles in

24. Any of numerous complex proteins that are produced by living cells and [bring about] specific biochemical reactions at body temperatures. See Notes for citation.

the entire universe—not atoms, but subatomic particles. The *difference* between 10^{80} and $10^{40,000}$ is a 1 with 39,920 zeros after it. If I were to show that number here, it would be a 1 followed by about 20 straight pages of zeros times greater than the number of all the subatomic particles in the universe. That's the odds against getting all these enzymes by chance. As far as I'm concerned, thinking that life somehow arose by chance from some primordial soup is completely unrealistic.

Further, in regard to "settled science," there's no such thing as settled science. Fairly settled, yes. But totally settled, no. Science can only infer. It can never absolutely prove something. Basically, what science does is advance a theory that best explains what we're seeing, then continues to use that theory until such time as it's disproven or a better theory comes along.

Personally, I believe the idea that matter can think is untenable. But the idea that something nonmaterial can think is really hard to deal with. What would such a thing be like?

Well, it would be nonmaterial. But what could possibly exist, yet be nonmaterial? Actually, lots of things are nonmaterial. Like beauty. Like ideas. Like respect or love or intention. These are all nonmaterial but exist "within" and impinge upon the physical.

I think we can develop a theory for life and this universe that fits the facts better, follows the rules of logic more closely, and explains more things than the view of life evolving from a bowl of organic soup.

Having said that, how do we get there from here? Well, to start with, "I think, therefore I am" could be seen as a proof of a more fundamental statement: "I am."

With that, we finally have an anchor on which to hang a philosophy of life and the actions undertaken by life.

I am.

I am here. I am a manifestation of life or living; I am alive. Knowing this, we might not be able to work out all the questions of metaphysics, but we can certainly use it, along with other data, to work out some of the questions of life as they pertain to this life, this place, this time.

CHAPTER 8

TOOLS OF LOGIC

How often have I said to you that when you have eliminated the impossible, whatever remains, however improbable, must be the truth?
—Sherlock Holmes (Sir Arthur Conan Doyle)

It's pretty clear that relying on physical-universe data alone will get us just so far. Somehow, we have to work out how to get trustworthy answers to important questions that require more than the solid physical data of the hard sciences.

That, however, is not something current Western culture is really much good at.

Depending on physical data exclusively solved a big problem. There was too much corrupt data, too many slanted observations, getting into the mix, and a way had to be found to get some discipline into our data. So, we chose empiricism.[25] Up until now, empiricism has worked really well for us. But it may well be that we're reaching the limits of what we can accomplish with strictly physical data, and so we need to pick up the question again of, "How do we get trustworthy data?"

Well, we have to learn, or relearn, a few things. For starters, we have to learn to be unswervingly honest, especially with ourselves. We have to be

25. The doctrine that all knowledge is derived from sense experience. See Notes for citation.

able to trust ourselves. We have to learn to look beyond the easy, apparent, or desired and see what's actually there.

If we can do that, we'll find that Western culture does have access to some workable answers. Good and usable tools for thinking already exist, some of which have been around for a long time and have had illustrious careers. They might have fallen on hard times recently, but that doesn't make them any less usable. We can take these tools, dust them off, combine them with data verified by experiment, with mathematical logic, and use this combination to sort out the best possible answer to the mysteries and paradoxes that plague both the hard and the human sciences.

So, what are these tools?

Conditionality—Scientific conclusions are at their root always conditional and subject to correction, even though we accept them as true for now. We make the best explanation we can with the data we have, and proceed to use that explanation until such time as new data becomes available that shows us that the explanation we've been using no longer holds water. If that occurs, we revise our theory with a new explanation that encompasses the new data.

Generality of causes—Basically, this one says that if you have a set of effects that are all the same effect, you can assume that they all had the same cause, even if they occur in different places or times. For example, rain, whether in North America or Africa, always has the same set of causes. The same effects are always caused by the same causes.

Universality of causes—If something is shown to be a general cause here (on Earth), we can take it to be a general cause anywhere and everywhere in this universe. Earth is not a special case, at least not in terms of universal laws. The laws of this universe manifest themselves here on Earth, and once we discover them here, we can use them anywhere.

Causal adequacy—The explanation we're coming up with should actually be reasonably capable of producing the effect in question; and the more likely it is that the cause in question might have created the effect we're seeing, the more likely it is that our explanation is correct. For example, if we see that a doll has moved from Point A to Point B, it's much more reasonable to assume that the doll's owner moved it than it is to assume that the doll got up and walked over there while we weren't looking.

Workability—The proper way to value something in this universe is by workability—how well a thing works to bring about a particular result under a set of circumstances, and how big that result is. Things that are generally true will work well, and truer things work better than falser things.

A datum has value in this universe to the degree that it is workable in producing a desired result under a given set of circumstances. Broader causes work to create bigger effects and thus have more value than limited causes. I should probably also note that value, as used here, does not mean "goodness." All it means is how big an effect gets created. That effect could be good or bad. The value could be positive or negative. As an example, a nuclear warhead is an example of a really workable bad effect. The data that go into that nuclear warhead have high value, but used the way they are, it's a high negative value.

Degrees of truth—Classical logic starts by viewing a proposition as either true or false. Once that determination is made, you use that datum to figure out the truth of other related data. A great example is, "If A=B is true and B=C is true, then A=C is true. Or, if A=B is false and B=C is true, then A=C is false."

Using "true/false" logic works fine for ideas, abstractions, and generalizations. It works great, for example, for many types of math.

However, logic as it applies to real life needs to come up with answers that are more subtle than just true or false, right or wrong. Real life is more complex than that. It's hard to make a picture using just absolute black and stark white. You need gray to make it look realistic.

After all, what datum in real life is totally 100 percent true with no moderating factors? What decision can someone make that's totally 100 percent correct under all conceivable circumstances? What thing in this universe can be held up as an example of perfection? In this universe, in these three dimensions, we can get close, but we can never get an absolute of any one thing.

No single thing in this universe is ever 100 percent true or false, right or wrong, because the interplay of opposites is at the heart of everything we can see or touch. It's always a combination, an interplay of at least one set of two opposing principles or absolutes, which combine to produce a real item or condition. The absolutes themselves are unattainable in physical reality. For example, there's no such thing in this universe as complete "no motion." One way of looking at it is to consider that the absolutes are nonphysical ideals that can combine to form physical reality. Combine "no motion" and "all motion" and you get real motion.

Simplicity—There's a tool used in reasoning called the *lex parsimoniae*, which can also be known as "Occam's razor."[26] The Latin means nothing more than "the law of simplicity," and it's pronounced "lex parse-ammonia." The Random House Kernerman *Webster's College Dictionary* says this is "the principle in philosophy and science that . . . the simplest

---

26. William of Ockham (c. 1287–1347) was an English Franciscan friar and . . . philosopher, commonly known for Occam's razor, the . . . principle that bears his name, and also produced significant works on logic, physics and theology. See Notes for citation.

of several hypotheses[27] is always the best in accounting for unexplained facts." Briefly stated, the simplest explanation is best.

This rule has been used in lots of different ways for centuries and is generally recognized as having value. But why does it always seem to work out that the simplest explanation is the best? Is it just some sort of lucky coincidence?

Let's refer back to the idea of a pyramid of knowledge. Any system, whether it be a physical system or a system of ideas, is built up starting with a few basic and fundamental data, pieces, or ideas; and these fundamental pieces are combined and recombined in increasingly complex ways until you get a whole huge system, which is filled with complex structures. Use matter in this universe as an example. There are just a few fundamental physical particles, yet they combine and recombine in increasingly complex ways until they produce all the marvelous and endless complexity of matter that unfolds before us. A few basics create all the wonderful complexity we see.

The same goes for ideas. Basic ideas (they can be called laws, principles, or axioms), start from a point of fundamental simplicity and combine together to form a pyramid of increasing size, complexity, and diversity.

Starting from things you already know, you can look "up" toward the point of the pyramid. If you do that, you'll be figuring out increasingly basic principles, and each new level of "basic-ness" will be able to explain more things. Knowing that different sorts of matter can combine to make new types of matter explains a lot of things. Knowing about molecules

27. A *hypothesis* (plural: *hypotheses*) is a proposed explanation for a phenomenon. See Notes for citation.

explains more. Knowing about atoms explains even more. Knowing about subatomic particles explains more yet.

Keep doing this, and eventually you'll get to the top of the pyramid. That's where you'll find the most basic principles that can explain everything else in that pyramid. One of the ways you can tell that you haven't gotten to the top of the pyramid yet is the existence of phenomena that your existing principles can't easily explain, or where you have to "add on" exceptions or special rules to your existing theories to make it all hang together. An example of this is the way fifteenth-century astronomers working with a geocentric model[28] had to tack on circles within circles to account for the observed motion of Mars.

So, the law of simplicity is not just a coincidence or a convenience. It's a statement based on how this universe is put together and how it works. Of all the tools given, this is probably the most important because it's the one that points to the top of the pyramid—the most basic truths.

Summary

Here we have a stable of tools that allow us to evaluate any idea or potential explanation that comes up for review.

- Whatever workable explanation we have now is fine and usable, but it's only good until such time as we come up with a better one.
- Are similar effects also caused the same way, or are we creating "special cases"?

---

28. According to geocentric model in use at the time, Earth is positioned at the center of the universe, and all celestial bodies, including the sun, moon, planets, and stars, revolve around it in perfect circular orbits. To account for irregularities in the apparent motion of planets, this model introduced the concept of smaller circles within the main circular orbits. See Notes for citation.

- Can we see that any effect like this, occurring anywhere, could have the same cause?

- Is our explanation reasonably able to cause what we're seeing without having to resort to too many highly improbable assumptions?

- If we put our explanation into effect, does it work?

- How well does it work? Is it elegant? Does it cover all the bases?

- Are there better or more reasonable explanations, or explanations that work better?

- Finally, and probably most importantly, given that there's usually a bunch of different ways of looking at something, which explanation requires the fewest outrageous or fantastic assumptions to achieve the greatest depth of explanation?

Using these, we can adapt Holmes's statement for our purposes here, making it, "When confronted by a series of highly improbable explanations for some phenomena, select the one which is the least improbable." Using this, we can reasonably evaluate any potential explanation (hypothesis) and pick the one that fits best, then apply it going forward until such time as a better one comes along. We can use these tools to arrive at answers to questions that cannot be answered by physical data alone.

The law of simplicity and these other tools of logic should be applied to the unusual experimental data from atomic physics and the developments in math to see if there could be an explanation for these strange phenomena that is simpler, more explanatory, more fundamental, and more comprehensive than any that we, as a civilization, have had so far.

Basically, looking at all the new data and knowledge we have, what's the shortest line and most effective line that connects all those points?

But before we start with that, we have a few more issues to sort out. Like . . . what is truth?

CHAPTER 9

THE ROAD TO TRUTH

In questions of science, the authority of a thousand is not worth the humble reasoning of a single individual.
—Galileo Galilei

Is there such a thing as absolute truth? Well, . . . we know that we exist.

We can make a few assumptions once we accept the idea that somebody does indeed exist. If you were to believe that all you perceive—that others, all life, the world, and everything in it—was nothing more than an illusion, then sciences are pointless and you don't need this book. Close it up and move on to higher (or other) realms. But if you have some belief in the reality of what you see around you; if you feel the floor beneath your feet, can hear the birds sing, and fruit tastes sweet to you; then you accept that the physical world and all in it have some actuality.

We exist, and this universe is actual enough.

Human sciences are areas of study and organizations of truths in this reality, this universe. We don't have to know or assert anything about the ultimate truth of universes or the ultimate nature of reality; all we need to do is get on with the business of handling this universe and this reality.

What will be our yardstick for evaluating truth in this universe? How will we know which data have value and which are false or misleading? We already know one thing for sure that does not need such an evaluation. We know that we are. You are. We all are, and we do think, and we know from this that we are alive, or manifestations of life. We accept that we all live in a universe that is real enough. We also know that we should evaluate data based on workability. But how do we work out the truths of this universe?

For part of that answer, we can look to Francis Bacon, a statesman, philosopher, scientist, and author who lived in the seventeenth century. His work established a new method for scientific inquiry, often called the scientific method. The modern scientific method is that single philosophical advance most responsible for humanity's dominion over the planet for the past five hundred years.

How does the scientific method help us sort out truths and the value of those truths of this universe?

First off, it's essential to look with the purpose of understanding, to look without bias or preconception, and that can be a lot harder than it seems. Evaluating data without bias or preconception requires strict self-discipline. However, our ability to do things and to understand the world about us is only as good as the clarity and quality of our observations of it; so, using self-discipline to make sure we're looking without bias or preconception is essential. The usefulness of our sciences will only be as good as our willingness and ability to make accurate and unbiased observations of *all* the things available to be seen.

After making observations, you evaluate what you've got, work through it, and throw a possible explanation out there. Then test that explanation and repeat as needed until you get some workable basic principles, then test them and refine as needed. Being honest with yourself is essential here. You have to use all the available data, not just the ones you like or are comfortable with. If your observations are good and you do this right, the eventual end results are the most basic principles of a field, its axioms, which both explain and predict everything in that field.

As we know, the value of any datum is the degree to which it matches a cause to an effect, and the size of the effect created. The value of a datum is directly related to its workability. Knowing with certainty how to cure a disease is more valuable than having some vague ideas about it, and knowing how to cure a disease is more valuable than knowing how to fix a floor.

Data that explains more things are more valuable because they can ultimately be used to create bigger effects. This is why basic laws are more valuable than data about specific or limited causes. Knowing and being able to apply the laws of motion is more valuable than just knowing about trajectories. One allows you to shoot guns accurately; the other allows you to handle any and all motion, anywhere. So, knowing the axioms, the most basic truths of a field, would thus be extremely valuable.

What if, in working this out, we find little bits of data that don't quite conform to existing theories? If 99 percent of the phenomena in an area are well explained by the current wisdom, shouldn't that be good enough?

It was the disturbing implications of a few odd fossils that led to the discipline of archaeology. And it was a few strange results from some odd experiments that led to modern atomic physics.

Rather than tolerating the little bits of data that stick out from under the rug of current theory, we should pursue them. We should pursue them regardless of any upheaval which that pursuit might create in the existing order of things, because that pursuit is what opens the door to new and better understandings of reality.

Our yardstick for truth exists independently of, and senior to, any existing preconception, viewpoint, or favored notion about how things work or how things "should" work. Truth is a stark and exacting master and will brook[29] no fables. This universe is what it is, regardless of our current preconceptions or preferences. We succeed as a race when some of us are bold enough to see what they see and hear what they hear; when they perceive the patterns of existence spread out before them, no matter how this perception might contradict some belief system spoon-fed since birth. As has been true in the past, so it will continue to be true; we will succeed to the degree we are able to live by this creed,[30] and fail to the degree that we shrink from it.

Knowing that we exist and accepting this universe as real, we can use Bacon's method to nail down some workable truths. However, this method will work only as well as we do. It depends on us for accurate observations, honest evaluations, and a devotion to truth. If we do those things, we can resolve many of the mysteries of life and this universe and use this data to make our lives better.

29. To bear; suffer, tolerate. See Notes for citation.

30. Any statement or system of beliefs or principles. See Notes for citation.

CHAPTER 10

MATERIALISM

The positivists[31] have a simple solution: the world must be divided into that which we can say clearly and the rest, which we had better pass over in silence. But can anyone conceive of a more pointless philosophy, seeing that what we can say clearly amounts to next to nothing? If we omitted all that is unclear, we would probably be left completely uninteresting and trivial tautologies.[32]

—Werner Heisenberg

It has become appallingly obvious that our technology has exceeded our humanity.

—Albert Einstein

Our current models of reality are all, in some way, flawed. The existing paradigms are almost all underpinned by a basic outlook on life and this universe, which is showing some serious cracks and is ready for an upgrade. What is this outlook, its strengths and flaws?

Materialism has been the default answer for our sciences and culture for a while now, and although it has produced some marvelous results, it has some serious shortcomings, not least of which is that it can't explain all

31. Positivism: a theory that . . . positive knowledge is based on natural phenomena and their properties and relations as verified by the . . . sciences. See Notes for citation.

32. A statement that is always true, . . . as in "either the sun is out or the sun is not out." See Notes for citation.

the new data that has come out of physics and math. It also leaves us high and dry when it comes to answering questions about ourselves and our societies.

Materialism

What is materialism? According to Merriam-Webster's online dictionary, it's the "theory that physical matter is the only or fundamental reality and that all being and processes and phenomena can be explained as manifestations or results of matter." If you're a materialist, the only thing that matters is matter, the energy that moves it, and the space and time it sits in.

Common sense will tell you a lot about how things are. There are such things as matter and forces, and they interact in certain ways. Newton's laws of motion, along with other physical laws, summed these up and arranged them so that they worked really well in predicting how things moved and acted. A large part of the modern world is built on these laws. If you want to build a rocket that will make it into outer space, you use them to figure how fast it will need to go, how much fuel you'll need to have to get it going that fast, etc. These laws allow us to calculate and predict the orbits of the planets, as well as giving us almost all the data needed to handle material objects here on Earth.

That view of how this universe works is called "classical" physics, and it still works really well for anything above an atomic scale. There are certain ideas embedded in the classical view of the universe:

- Matter, energy, space, and time exist.
- They exist independent of us.
- We somehow came into being in this preexisting universe.

- We have since learned (and continue to learn) how to manipulate it to suit our needs.

There were lots of new ideas that popped up as a result of the discovery and use of classical physics. One was that energy (and by extension, matter) could not be created or destroyed but only converted from one form to another. If that was true, then if you knew all about a mass and its motion and the forces related to it, you could predict with precision how that mass was going to move in the future. You could also predict how it would react to other forces.

Looking further, this meant that if you were able to know all the positions, magnitudes, and directions of all the objects and forces that existed in this universe at any moment, you would be able to predict the entire future of all objects and forces.

This meant that, once energy and mass got moving in this universe, all the motions and interactions would spread out from that starting point in a totally predictable manner.

Like a machine.

That gave birth to the view of the universe being a sort of "clockwork." It got wound up once upon a time and has been unwinding in an unchangeable manner ever since. According to one version of this idea, we're part of the machinery, and our actions were all predetermined back when the universe was "wound up." Power of choice is an illusion. Everything was already fated to occur.

The debate about whether or not we have any power of choice in our actions is still in progress. But what if you don't fully subscribe to the "it's already all fated" idea? It still seems that the universe is what it is, that it *somehow* exists, and that we exist within that universe. Matter and energy simply are and always will be. While we might have free will to move STUFF around, the STUFF that we're moving exists objectively and independently of us, and all of our lives and actions are played out against a background that existed before we ever came on the scene and will continue to exist long after we leave.

In fact, we ourselves are composed of nothing more than the matter and energy that was already present. Life *somehow* arose out of this STUFF, *somehow* evolved from totally simple to extremely complex forms, then *somehow* developed intelligence, which *somehow* became self-awareness to go with the intelligence. As this story goes, a long series of incredibly lucky evolutionary breaks led us to live better lives as, essentially, smarter animals.

Scientific materialists have tried very hard to show that this universe does not have a beginning, but they're having a tough time of it. Recent scientific work, especially the (original) work of Hawking,[33] Ellis,[34] and Penrose,[35] strongly implies that the material universe started, and started from nothing, a long time ago—at least nothing spatial, temporal, material, or physical.

Similar issues arise when trying to show that, somehow, the chemical basis of life got started by some sort of chance or mechanical means. Asserting that the chemical basis of life got started in this way is akin to saying that this book came together due to the specific chemical attractions between

---

33. Stephen Hawking, 1942–2018, English theoretical physicist, cosmologist, and author. He was the Lucasian Professor of Mathematics at Cambridge, widely viewed as one of the most prestigious academic posts in the world. See Notes for citation.

34. John Ellis, born 1946, British Swiss theoretical physicist. See Notes for citation.

35. Sir Roger Penrose, born 1931, British mathematician, mathematical physicist, philosopher of science, and Nobel Laureate in Physics. See Notes for citation.

ink and paper. Attractions between ink and paper might yield blots, but you would never get intelligible words, much less a book.

The odds of a single operational protein arising by chance from a "primordial soup" are about 1 in 10^{164}. That's 10 with 164 zeros after it.

Furthermore, those are the odds of generating *one* single, solitary operational protein. Every cell in our body is made up of thousands of these proteins, and they all have to operate together in an intricate and complex dance in order for that cell to survive. In regard to that cooperation, the cells of our bodies could be thought of as being similar to an automobile: not only are there thousands of complex parts, but they have to be the right parts in order to work together properly. Toyota parts, while they might be lovely and precise, don't work in Fiats. You can't survive with just any proteins. They have to be the right ones, at the right time, in the right place.

In short, the idea that the chemical basis for a living cell somehow arose spontaneously from a primordial soup is akin to asserting that, if you dumped a few hundred tons of raw earth in a pile and waited long enough, a Ferrari would assemble itself out of the mud.

Going further still, let's say we concede that living cells came into existence from nonliving matter one way or another. How, then, do we explain the arrival of consciousness? Did that somehow self-assemble too?

Hmmmm . . . do you notice a lot of "somehows" here? When you're trying to explain how things work, having to rely on lots of "somehows" is a bad sign.

That philosophy, with all of its somehows, is materialism. It holds that STUFF (matter, energy, space, and time) is all there is; that we can know about the STUFF and how it works, but that, ultimately, we're just STUFF ourselves and our thoughts are nothing but the interactions of brain chemicals made up of STUFF. The universe exists whether we're participating or not. It's "out there" and will continue to exist unchanged after we're gone. We're no more than shimmering sparks adrift on an infinite sea, shining for a moment, then gone forever.

Materialism has given us a set of sciences, which have worked and given us a stability that was lacking after centuries of confusion and turmoil. A lot of that confusion was due to divergent religious dogmas, or different philosophies of life. Prior to the twentieth century, neither religion nor philosophy really got the job done, and this failure left a power vacuum into which stepped materialism with its repeatable results, its workability, and its solidity.

Unfortunately, science got "married" to materialism, and the combination was so successful that it led to a state where other ways of looking at things were crowded off the stage and scorned as being "unscientific." After all (according to its proponents), materialism invariably worked; it got us rich beyond the dreams of ancient kings; got us healthy enough so that many of us could live out what was thought to be our full life spans; and even got us (briefly) off this planet. Religion and philosophy seemingly only got us a bunch of questionable doctrines, fuzzy belief systems, religious wars, crazy ideas, and New Age touchy-feely junk.

Materialism was so successful that it has given us a sort of mental paralysis. Its basic concepts have been so ingrained into our consciousness that we tend to see the universe of masses and energy as the only real truth. Thoughts, emotions, aesthetics, and other similar things are seen as only the fleeting and flickering interactions of brain chemicals in an animal called Man. Having had their day, they will come and go, leaving an unchanged and uncaring universe behind.

Materialistic philosophy, while having many strengths, has some serious shortcomings too. It can't explain certain vitally important subatomic phenomena. Not only that, while we can split the atom and send people to the moon, we can't seem to sort out how to reliably educate our kids or eliminate poverty, stem the tide of dissatisfaction and depression, or figure out how to handle wanton violence, to mention just a few things. While materialism has given us the ability to destroy our world, it has not given us the sense to be able to know how to build one worth living in.

However notable its shortcomings, materialism was the predominant way of looking at things as the twenty-first century began. As workable as it may have been, there was some fraying at the corners of this fabric, and not just on a societal level. There were certain scientific experiments done by materialistic scientists according to strict materialistic criteria, but that gave results that could not be easily reconciled with materialism. These experiments produced confusion, mysteries, and paradoxes. The results of these experiments are not in doubt, have been subjected to the strictest scrutiny, and have been validated. What has remained in doubt is what those results mean or how to explain them. How do we resolve the mystery?

Materialism has brought us a long way, but recent data has shown that there are mysteries that it simply cannot explain and desirable things that it cannot do. Anytime we have an outlook on life or a school of thought that is clearly starting to fail, that means it's time to start looking for a better system, a new outlook that will embrace all the data the old system can't digest, explain what is currently inexplicable, and, in doing so, open doors currently closed to us.

Nevertheless, before we go jumping off toward answering these questions, we owe it to ourselves and others to learn a bit about this universe, how it works, and exactly how or where it is that materialism has come up short. Let's find out how STUFF works.

CHAPTER 11

MATTER AND ENERGY

In order to get a grip on how STUFF works, we'll need a bit of background. First, we'll hit some basic principles based on current science and theory. If you already know all this, great! This material has been covered thoroughly in all sorts of different books and lectures. But not everybody knows these things. These chapters provide an overview so that those of you without a science background will be able to get the data you need to know without needing a doctorate degree. On the other hand, if you truly don't care how STUFF works and just want to get to the heart of the matter, I suppose you could skip over chapters 11 and 12 and go straight to chapter 13.

Matter

Matter is composed of atoms. There are over one hundred different kinds of atoms currently on Earth, ninety-two of which occur naturally (the rest are man-made). Atoms themselves are made up of smaller pieces, which are called protons, neutrons, and electrons. Protons have a small positive electrical charge, electrons have a similarly small negative charge, and neutrons have no particular charge. Protons and neutrons are about the same size and mass and generally form a tight and stable core (called the nucleus) for any atom, while electrons are much smaller and lighter and "fly" around the nucleus. In the past, we thought of these things as a little solar system, with the protons and neutrons at the core, and the electrons orbiting around that core.

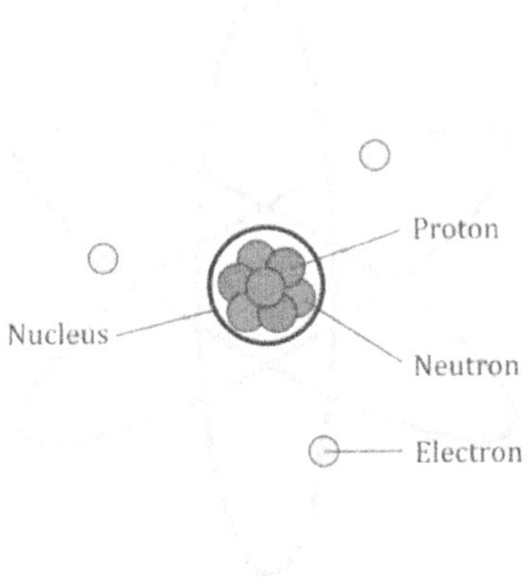

Figure 1: Solar system model of the atom.
Source: AG Caesar – Wikimedia Commons, CC BY-SA.

These protons, neutrons, and electrons tend to settle down into stable patterns with equal numbers of protons and electrons. Atoms of a given number of protons (for example, two) all act pretty much the same. In fact, each atom with two protons is considered to be a particular type of matter. An atom with two protons (and usually two electrons) is called helium.

Hydrogen has one proton and one electron; helium has two of each; carbon has six of each; oxygen has eight of each, and so on. All atoms with six protons and electrons are always carbon. An atom with eight of each is always oxygen, and so on. These different types of atoms are called elements. The word *element* comes from a Latin word meaning "basic principle." Other examples of elements are sodium, chlorine, iron, gold, silver, potassium, and neon.

Elements can get "married"; i.e., they can get bound together to form compounds. A compound is a substance that is made up of two or more atoms bound (more or less) stably together. When you get a couple (or more) atoms bound together, they form a molecule. Most of the things we encounter on a daily basis are compounds. Examples of compounds are things like sodium chloride (sodium and chlorine = salt), iron oxide (iron and oxygen = rust), dihydrogen monoxide (hydrogen and oxygen = water), and carbon dioxide (carbon and oxygen). Some molecules, especially in living organisms, can get really big and really complex. DNA is a good example of a really big and complex molecule.

Even though the protons and neutrons generally form a stable nucleus, an atom can, depending on the circumstances, sometimes pick up or lose an electron or two, or even share electrons with other atoms. That's frequently how atoms get "stuck" to each other to form compounds. Adding or subtracting neutrons or electrons from "normal" matter can also form different and interesting subforms of matter.

Energy

In addition to matter, there's something called energy. Energy shows up in or as motion, or potential motion. Motion occurs in or through space and takes time. Energy can be seen as flowing or spreading through space, although it can also get condensed or solidified. When it gets condensed, it's called matter. Ever since Einstein published his famous equation $E=mc^2$, scientists have viewed matter as being a sort of condensed energy.

All Einstein's equation means, by the way, is that the amount of energy in a chunk of matter equals the amount of matter times 186,000, and then that number gets multiplied by 186,000. If you multiply 186,000 times 186,000, times some other number, you'll get a really big number. There's LOTS AND LOTS of energy in matter. In fact, the energy released in an atomic-bomb blast is only a small amount of the total energy present in a few pounds of matter.

Energy travels in waves. These waves are similar to the waves that you would get from dropping a rock into a calm swimming pool, or maybe the waves you would see if two people held both ends of a rope and one of those two started to flip one end of the rope up and down. The waves have crests and troughs, which can occur in a regular pattern. The distance between the crests is called wavelength. The waves in the water (if viewed from the side) or the waves in the rope would look like this illustration.

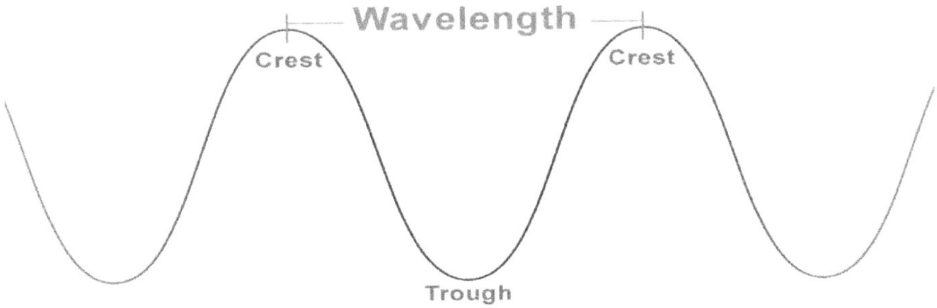

Figure 2: Wavelength crest and trough.
Source: Geoff Ruth – Creative Commons.

Light is a form of energy and was shown in the early 1800s to travel in waves. In dealing with visible light, different wavelengths correspond to different colors. Generally speaking, flows of energy, such as sound, heat, light, or x-rays all travel as waves. Different wavelengths for these things can bring about different characteristics. For example, as the wavelength of light gets shorter and shorter, it goes from red light, to blue light, to ultraviolet light, to x-rays, to cosmic rays. Wavelengths can be detected and measured.

Electricity is also a flow of energy. One theory is that electricity is a disturbance of the electrons around an atom, and that disturbance travels from atom to atom through a substance. It would be something like watching people doing "the wave" at a big stadium.

So, matter is made up of different combinations of little particles, which come together to form atoms, which then bind together to form molecules, and so on, to form everything we see around us. Energy is motion going through space, which takes time. Energy travels in waves. As of the end of the nineteenth century, this (along with the principles of electromagnetism[36]) explained almost everything there was to know about how this universe operates. There were just a couple of loose ends. . . . For example, nobody could really explain why shining light on certain things generated electricity. Also, what was it that made us sure that energy travels in waves?

Let's get some more data about these loose ends.

36. The phenomena associated with electric and magnetic fields . . . and with electric charges and currents. See Notes for citation.

CHAPTER 12

BENDING ENERGY

How did we know that light travels in waves? There was an experiment done by Thomas Young[37] back in the eighteenth century that showed conclusively that light had the properties of waves. Given that, what wave properties are we talking about?

You can do an experiment with waves in water and a wall with a gap in it. Looking at a gap in a seawall works if you'd like. Make it so that the distance between waves in the water and the gap in the wall are of a similar order of magnitude. It's not like the distance between waves should be inches while the gap in the wall is miles wide.

Let's say that the wavelength (the distance between tops) of the waves is 20 feet. Let's also say that the gap in the wall is 100 feet, much longer than the wavelength. What do you see when the waves pass through the gap in the wall? If the waves are moving straight ahead when they hit the wall (and the opening), that part of the wave that hits the wall just bounces off, while the part of the wave that goes through the gap goes straight through and carries on. So, you get a wave as wide as the gap in the wall moving straight toward shore. No surprise there.

However, an odd thing happens if you make the gap in the wall smaller than the wavelength. Let's say you still have waves with a wavelength of 20 feet, but you make the gap in the wall 18 feet. You'll see that the waves bend a little bit on the other side of the gap. If you keep the waves at 20

---

37. British scientist, mathematician, physician, and linguist, 1773–1829. See Notes for citation.

feet but make the gap 5 feet, they bend a lot more on the other side; and if the gap is only 1 foot, the bending really gets to be substantial. That bending effect is called diffraction. Here's a diagram of what this might look like if you were looking down onto the waves from above:

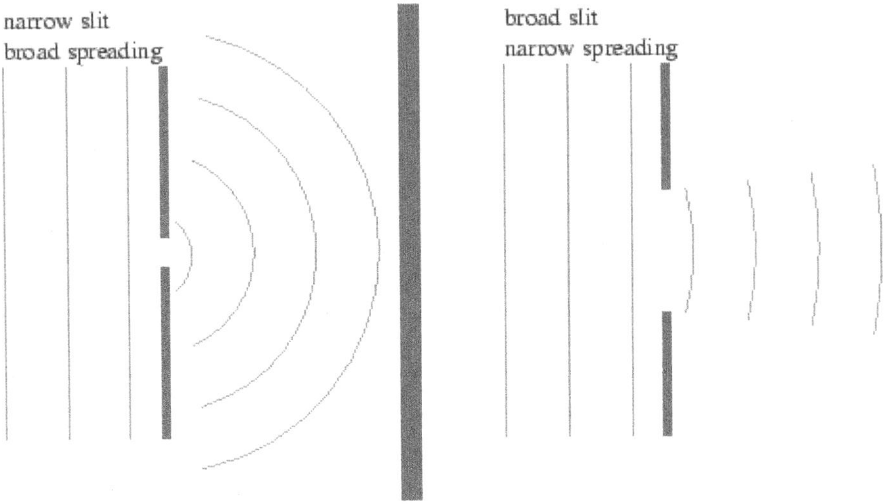

Figure 3: Examples of diffraction.
Source: Jordgette – Wikimedia Commons.

This same bending effect applies to all waves. That's why you can hear somebody talking around a corner but can't see them. Sound waves are MUCH longer than light waves (millions of times longer, as a matter of fact) and bend a lot when they hit a gap like a doorway. So, the very long sound waves actually bend around through the open doorway, while the much shorter light waves just go straight and don't bend at all. Thus, you can hear somebody around the corner, but you can't see them. The waves don't change in size or wavelength, but they can change direction.

What happens if you have two gaps in the wall, fairly close to each other, and both gaps are small compared to the wavelength of what's passing through them? The waves going through both gaps will bend and fan out on the other side of the wall. However, these sets of waves from each gap will eventually encounter each other on the other side. When they do, they will interact. If you think of these waves as being waves in water, some of the high spots (the wave peaks) from one gap will hit the high spots from the other gap, and they'll stack up on, or reinforce, each other to make a "super-peak." Some high spots will hit low spots (troughs) and cancel each other out, making a dead spot. A pattern will form like this:

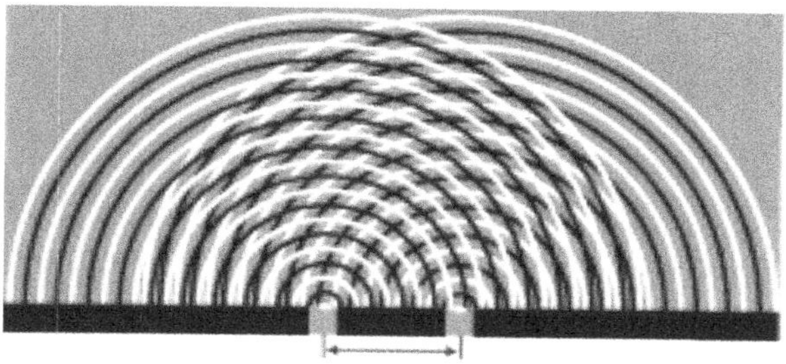

Figure 4: Double-slit diffraction. Source: Bcrowell – Creative Commons.

The waves interacting with each other is called interference. The places where one peak sits on top of another peak are particularly high. In water, one of the 2-foot peaks gets added to another 2-foot peak, and you get a 4-foot peak. This is called constructive interference. Where you get the meeting of a peak and a trough, the 2-foot peak gets canceled out by the 2-foot trough, and the water level in that particular spot isn't any higher or lower than normal; it's a dead spot. This is called destructive interference, because a trough is "destroying" a peak (and vice versa).

That's how waves in water build up and can get really big. Go out sometime and watch wind blowing across still water. It stirs up ripples, and soon some of the ripples interact with other ripples, and you'll get spots where the ripples add to each other and get "supersized." Eventually, if this continues, the "super" ripples add up to waves, and the waves can get bigger and bigger, especially if the wind keeps blowing hard. The same thing occurs with any kind of waves.

So, we've got two different characteristics of waves: diffraction (curving after going through a relatively small opening) and interference (the interaction of one set of waves with another). All waves have these characteristics; anything that moves in wave form will do both of these things. However, things that don't move as waves will not. Therefore, light waves going through a small hole will diffract on the other side, sound waves will interfere with each other, but a ball thrown at a hole in the wall smaller than it is just bounces off.

An Experiment—Bending Light

That's all fine, but let's go back to the question asked initially. How do we know that light travels in waves?

Thomas Young did a simple experiment. He had a light source that shined at a barrier that had a really narrow slit cut in it. Beyond that, he had a screen to see the light that passed through the slit. The slit was narrow enough that it was on the same order of magnitude as the wavelength of light—it was about the width of a razor slice. What did he see? He saw a small blob of light on the screen, much wider than the slit in the barrier. He could see that the light, instead of moving through and past the slit in a perfectly straight line and forming a razor-sharp line on the screen, was curving and spreading out on the screen, making a blob. It was diffracting.

That was surprising enough, but he went even further. He made a second narrow slit in the barrier near the first one and shined the light on the barrier again. But this time, what he saw on the screen was not two blobs like he expected, or even just a big, wide blob. Instead, it was a set of alternating dark lines and light lines in a pattern.

Apparently, the waves of light that were coming through each slit were diffracting, then interfering. They were fanning out, and then these fanned-out patterns interfered with each other, forming a pattern of alternating bright lines and dark lines. These bright and dark lines corresponded to the "super-peaks" and dead spots of the waves we saw in water, where the bright lines were the "super-peaks" and the dark lines were the dead spots. Here's a diagram of what he saw:

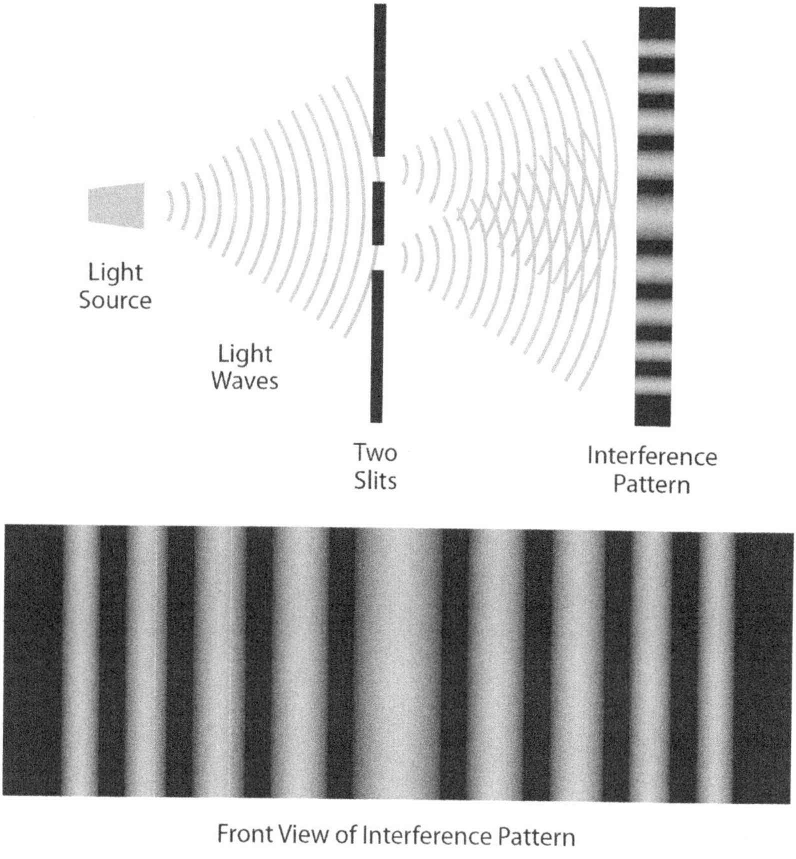

Figure 5: Interference causing a visible interference pattern.
Source: William Truax.

The interaction of the light waves on the far side of the barrier created the pattern of alternating bright lines and dark lines shown above. It's called an interference pattern. This was, and is, accepted as evidence that light travels as waves. It worked really well to explain and predict all sorts of things and was how everybody thought about anything that had wave characteristics for a couple of centuries, until the early 1900s. Here's an image showing what the blob (from the single slit) and the interference pattern (from the double slit) from a real experiment actually look like:

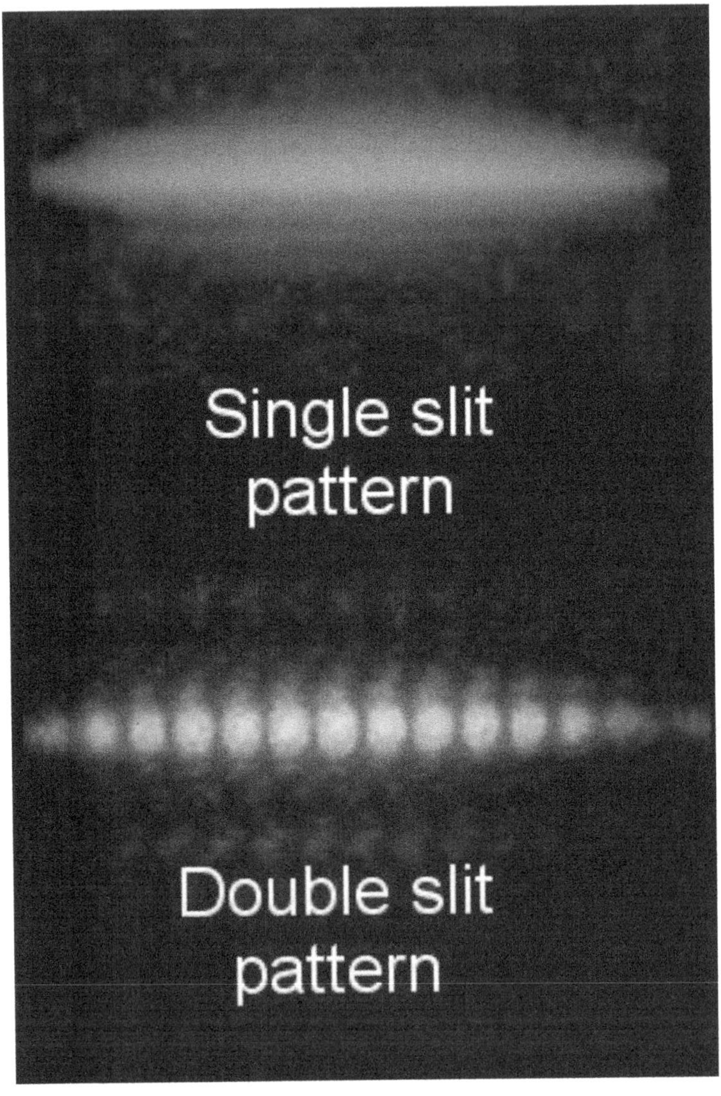

Figure 6: Single and double slit interference patterns.
Source: Epzcaw – Creative Commons.

So, waves have certain properties, and Thomas Young's early nineteenth-century experiment (called the two-slit experiment) showed that light has these wave properties. He concluded that light must therefore be wavelike in nature, and everybody agreed.

All settled, right? Light is a wave. So, why is it that shining light on certain materials generates electricity? How and why does this happen? Nobody had an explanation for it until Einstein. What did he have to say about this?

CHAPTER 13

THE BEST MYSTERY OF THE TWENTIETH CENTURY

For some reason, shining light on some materials generates electricity. This is called the photoelectric effect. Nobody knew why it occurred, just that it did. Early in the twentieth century, Einstein explained the photoelectric effect by saying that light was a stream of very small particles, which he called photons. He said these particles were hitting the photoelectric material, causing electrons to be knocked loose and then flow away. The theory worked great, explained lots of effects and predicted others, all flawlessly. The only problem was that it did so by supposing that light was particles, not waves. So, which was it? Was light a wave, as had been proven centuries earlier, or was it somehow a stream of particles, like little golf balls or something?

The answer, which was worked out then, and is the most commonly accepted explanation today, is that . . . it's both. It's chunks and it's not chunks. Light travels as a wave, but once it interacts with anything, it becomes (acts as) a particle. So, you could say that a photon starts as a wave and ends as a particle. It depends on how you look at it and what you're looking for. While this may not be the most satisfying answer, the math on it works wonderfully and has enabled many of the huge advances in technology that have occurred over the last sixty to seventy years.

A few years after Einstein's theory, somebody did the old two-slit experiment again, but instead of sending a stream of photons all at once, they sent only one photon at a time. However, even though they only sent one photon at a time, they were amazed to see that, over time, an interference pattern built up. This was a complete mystery. How could one light photon interfere with some other light photon that didn't exist?

This was confusing enough, right? But then, they also noticed that merely looking at a photon of light seemed to be the catalyst[38] for that light changing from being a wave to being a particle. Observing a light photon changed it from a wave to a thing. Why did merely looking cause this change? After all, waves and particles are really different things, and it seemed like light should be one or the other.

And then more experiments were done, this time with little bits of matter, not just energy, and they got the same results. Matter also had characteristics of both particles and waves and changed from one to the other when you looked at it.

This was confusing on several fronts. First, bits of both matter and energy seemed to be both waves and particles. Next, they both had the property, when a series of single bits (of either matter or energy) were sent into the two-slit experiment, of somehow interfering with other bits of matter or energy that had not yet been created or were already gone. Finally, the apparent solidness of both was influenced, maybe even controlled, by the decisions of the observing scientists.

What was going on here? Let's get more data. Let's recap the actual experiments.

The Experiments

Since the time of Young's two-slit experiment, technology had progressed so that scientists could work in the laboratory with one photon at a time instead of being locked into using streams of light. So, they did the two-

38. A person or thing that causes a change. See Notes for citation.

slit experiment again, but this time with one photon at a time instead of a stream of photons.

In the first stage, one photon at a time was sent at (and through) a barrier with a single slit. The photons landed on the screen behind the slit, and the position at which each photon landed was recorded. This was done lots of times, and after a while, a pattern started to show up on the screen. It was a blob pattern. The more photons that went through the slit, the better defined the pattern became. Since a beam of light is just a stream of photons, you would expect that a bunch of photons sent individually through a particular slit over a span of time would form the same pattern as a bunch of photons sent through that same slit all at once. The only difference was that, instead of getting the pattern right away, it took some time for the pattern to build up.

The next stage was a little different. In this stage, single photons were again sent at the barrier. But this time, both slits were open. Once again, a pattern gradually emerged on the screen behind the barrier. But instead of forming the two-blob pattern you might expect like this . . .

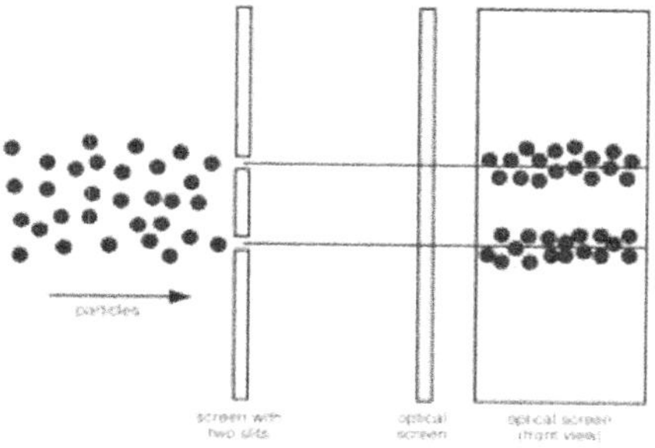

Figure 7: What you might expect with two slits.
Source: inductiveload – public domain.

the familiar pattern of alternating light and dark lines began to emerge. As the pattern formed, the successive images of the screen looked this way (each white dot is where a single photon hit the screen):

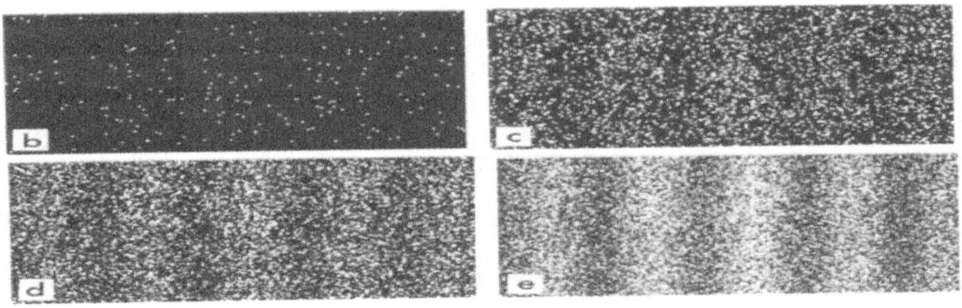

Figure 8: Interference pattern accumulating over time.
Source: Dr. Tonomura and Belsazar – Creative Commons.

It was an interference pattern. Not only that, if the second slit was closed, the interference immediately stopped, and the blob pattern began to form. If the second slit was later opened, the photons immediately began to form an interference pattern again. Opening or closing the second slit caused the interference to stop and start. When the second slit was open, the photons were behaving as waves and interfering with other photons, only they were interacting with other photons that did not currently exist.

The experimenters then tried putting a photon detector at various points along the path of the photons and noticed that if you figured out any sort of a way to observe, detect, or measure a photon in flight, or if it left any sort of observable record, anywhere along its flight path, the photons would instantaneously start to form the single-slit blob pattern.

That meant any sort of observation caused the photons to act as particles instead of waves. They also saw that the mere possibility of observation, or even an interaction with something else that was being observed, changed these photons. Observation, direct or indirect, changed what they were, where they appeared, and how they behaved. Unobserved, they behaved as waves. If observed, they became objects.

Adding to the mystery, more recent experiments have shown that this effect holds even if the experimenter doesn't decide whether or not to make an observation until after the photon passes through the slit(s). The experiment done was to send a photon at the slit(s), let it pass through, *then* decide whether or not to observe it. Even if the observe/no-observe decision wasn't made until after the photon passed through the slit(s), it still changed how the photon went through the slit. The observation decision changed the photon's *earlier* behavior as it passed through the slit(s). This decision to make an observation after the photon passed through the slit(s) created an effect that apparently went backward in time.

Think about these things for a moment. If an interference pattern is formed by the interaction of waves of light with each other, how could one possibly form when you only sent one photon at a time at the barrier? How could that one photon interact or interfere with another photon that was already gone, or wasn't even created yet? And why was it that simply looking at a photon, or even just being able to look at it, would change it from a wave to an object? And how could making the decision to look at it *after* it had passed the point when it should have turned into either a wave or object influence that "choice" retroactively?

And these waves that we've been speaking of—what were they waves of? We talk about these waves as if they were some actual thing. But there was *nothing* there to wave until some observation was made or could be made. What could possibly be waving?

When you sent a single photon at a single slit, it just went through. But if there were two slits of the correct size and location open, that single photon "knew" as it passed through one slit that the other slit was there, "knew" that the other slit was open, and "knew" enough to be influenced by it. Even worse, it seemed that a photon could "know" that it had been observed, could have been observed, or was going to be observed, and change its behavior accordingly.

Going still further, you can do a similar experiment with particles of matter instead of energy. Experiments of this sort done with electrons or other subatomic-level particles of matter get the same results. These fundamental chunks of matter not only can act like waves, showing diffraction and interference, but they seem to "know" what's going on around them. They even seem to know if they're being observed or not, or might be observed, and change their characteristics accordingly. Not only do the most basic bits of energy act in unexpected ways; so do the most basic bits of matter. Of course, if we consider that matter is a sort of condensed energy, it follows that the most basic forms of matter and energy would have some similar characteristics.

There are other experiments that show single photons that seem to "know" what happens elsewhere in space, or pairs of photons that seem to "know" what happens to each other, all instantaneously, with no time involved whatsoever, regardless of how far apart they are.

What is going on here??? Since when do subatomic-level bits of matter or energy know what is (or might be) going on around them and make decisions about where to be, what to be, and even whether to exist or not?

If you find these results confusing or inexplicable, just imagine how the scientists doing the experiments felt. How could you possibly explain these things? Yikes.

One thing the scientists noticed was that the issue of observation, even leaving any sort of a physical trace that could potentially be observed, was vitally important. Each photon that the experimenters sent through their apparatus acted as a wave until something happened that caused it to become observed, observable, or interact with an observable, at which point it became a particle—a thing. However, before that interaction occurred, it was *nothing*—no physical thing.

In these particular experiments, the screen that the photons were hitting made the photon observable. Until the photon interacted with that screen, the scientists couldn't tell what it was or where it was. It was just a gradually spreading zone of "potential particle," which obeyed all the laws of waves until an observation or interaction event, at which point it became a real particle. This led them to conclude that these waves were waves of potential—potential particles, spreading out indefinitely in time and space. There was nothing else "there." Prior to observation, there was nothing but a gradually spreading potential. Once you could (or did) observe it, this potential "became" a thing—it became a particle.

And what happens to our particle once observed? Does it stay a particle forever? Well, no. If observation and interaction with other observables cease, the particle gradually loses its solid reality and reverts back to a set of potentials.

As incredible as this all sounds, every bit of it is true. Of course, these are certainly not all the details about, nor a complete list of, all the experiments that show these effects. The information has been simplified. However, while it is simplified, it's still true. These effects do actually happen, and there's no disagreement about that from any top scientists. They know these effects occur. They just have trouble explaining them.

So, we have these unbelievable, but verified, experimental results. Subatomic particles seem to exist only as waves of *potential* matter or energy until they're observed, could be observed, or interact with something observable, at which time, and not before, they become real. If observation ceases, they gradually revert back to being potential again. In order to accomplish this, these particles seem to know what's going on around them and act accordingly.

This is one of the best mysteries of the twentieth century, and it hasn't really been solved yet. Let's get some more data that may help explain what's going on.

CHAPTER 14

POSITION AND SUPERPOSITION

Subatomic particles are the basic building blocks of this universe. They start out as nothing, then become something, but what they become is maybe this or maybe that, maybe here, maybe there. What and where they become depends on what's happening around them. How does a particle manage to be a "solid" thing, yet be elusive? It seems like it should be one thing or another, and not some odd combination of both.

What gives? How do they manage this juggling act?

It has to do with the actual nature of these particles: what they are and what they are not. The actuality of what they truly are is very different from what they seem to be.

As we found out in chapter 13, subatomic particles fundamentally exist in a state of potential, rather than reality. Potential is what you could call their "default" state. Coming into reality is a change from their default.

Another issue we came across in chapter 13 was the importance of observation. The scientists noticed that each photon they sent through their apparatus was not really "there" in any understandable sense until something happened that caused it to become observed, observable, or to interact with an observable.

This led them to conclude that these waves themselves were waves of potential, spreading out indefinitely in time and space. Prior to observation or interaction, there was nothing but a gradually spreading potential. Once you could (or did) observe it, this potential "became" a thing—it became a real particle.

Why is it that interaction with another particle causes a particle to actualize? One theory is that it's because interaction with another particle is ultimately an interaction with some observer, or something which is being, can be, will be, or has been observed by an observer. Joe looks at particle A, which bumps into particle B, which bumps into particle C, and so on. Regardless of the ultimate cause, particles don't actualize or "become" on their own. They need a catalyst.

Until that interaction occurs, those things we call subatomic particles exist in a state referred to as superposition. What superposition means in this usage is "beyond position." In this state, the particles don't exist as any recognizable thing. They have no particular location or characteristics—they have the potential of lots of different locations or characteristics, and they have all those potential locations or characteristics all at once. Using location as an example, they could be said to exist anywhere or everywhere within a given volume of space at any one time. Prior to interaction, any particle has a large set of potential locations and characteristics.

We can't really predict where a subatomic particle is going to show up upon interaction and what its characteristics will be when it does. What it is, where it is, or what it becomes is partially random and also depends partially on what we do and what we're looking for.

Interaction causes the particle to lose most of its potentials and actualize as mostly a single thing with particular properties, such as a particular location. We talk about this as the particle being or having become a particular thing—an object. However, even in this condition, the particle has not lost all of its potential properties. It is not *entirely* in one place. It has not *entirely* become a single thing. It still retains some very small quantity of other potentials.

If the interaction ceases, the particle afterwards reverts, gradually reassuming its original full set of potentials. It becomes less and less a single identifiable thing and more and more a set of potentials. This continues until the next interaction.

Subatomic particles generally interact with other subatomic particles countless times per second.

If they're anywhere near other particles or observers, this rapid and continuing interaction makes these subatomic potentials spend the vast majority of their existence looking mostly like particular things in particular locations with particular characteristics. As a result, they act and look like identifiable real objects to an observer.

Even though the basic particles and motions that make up our universe are unavoidably indeterminate and random in some way, normal-scale events that are very predictable come about by combining large numbers of random events. The world looks stable to us. The laws of motion work.

The only logical way to make sense of this is to recognize that we are, in some curious way, related to this universe. On a subatomic level, it's our decision about what to observe that determines what that thing we're observing is, that it is, where it is, and where it goes. The things that common sense tells us have a firm, separate, and objective existence, are instead potentially one thing or maybe another, and it is ultimately only our interaction with them that causes them to become those things that we then see. Until they interact with us, or interact with some particle that interacts with us in some way, they have no particular reality.

Once you look at all this data, sort through and make sense of it, you come to realize that these data constitute the death of materialism as a workable and comprehensive way of viewing life and this universe. Materialism was based on the idea that physical reality is the only reality, but this doesn't seem to be the case. The universe does not exist "out there," completely independent of us. Somehow, we're involved. As comfortable or comforting as materialism may seem, that's just not how things are. Materialism is an outmoded relic of our past.

As mind-bending as this is, you haven't heard it all yet. There's more to the puzzle. The next part? Just one of the most outrageous pieces of data to come out of modern math and science.

CHAPTER 15

INDETERMINACY[39]

Science cannot solve the ultimate mystery of nature. And that is because, in the last analysis, we ourselves are part of the mystery that we are trying to solve.

—Max Planck

"Uncertainty" is NOT "I don't know." It is "I can't know." "I am uncertain" does not mean "I could be certain."

—Werner Heisenberg

So, what's this hugely outrageous datum that came out of modern math and science? And how is knowing that going to help us?

The outrageous datum? At a subatomic level, there's a minimum amount of unknowability and unpredictability built right into the fabric of the universe.

We've already found out about particles that simultaneously have two different sides or states of being that are incompatible with each other. Scientists call this the principle of complementarity. This principle holds that *all* subatomic-level manifestations have two different and irreconcilable sides. One example of these different sides is waves and particles. They're completely different, but they're both there. As we saw

39. The condition or quality of being not precisely fixed in extent; indefinite; uncertain. See Notes for citation.

in chapter 14, a wave of potential becomes a particle, which can go back to being a wave of potential. This could leave us wondering—which is it? A wave or a particle? A thing or a potential thing? As we now know, it can be either at any time, depending. It's both.

Something else unpredictable shows up when you try to measure what's going on with these things. For example, experimenters can accurately measure either the position of a particle or its momentum (momentum is the speed, mass, and direction), but not both at the same time. To the degree they are accurate in determining position, the measurement of momentum becomes inaccurate. To the degree the momentum measurements are straightened out, the position gets fuzzy.

If the position is measured with complete accuracy, they can't measure anything at all in regard to momentum, and vice versa. If they try to measure both at the same time, the best result they can get is somewhat fuzzy measurements for both. It's like trying to tune into two nearby different radio stations at the same time. You can get one or the other clearly, or you can get fuzzy reception of both, but you can't get both clearly at the same time.

Not only that; to the degree they are unable to precisely measure either position or momentum, the outcome of that characteristic becomes random. For example, when you're looking at momentum, you're looking at where something is headed, how fast. Knowing that, you should be able to sort out where it's going to end up. However, to the degree they can't measure momentum, the final location of the particle cannot be determined in advance. Where it ends up is, to some degree, a matter of chance; although the odds of any one particular outcome can be known. Indeterminacy and randomness are directly related.

This proved to be deeply unsettling to a lot of people. If you throw a ball, you have a pretty good idea of where it is, what direction it's going, how fast it's going, and where it's going to end up. This was true in all cases in classical physics. So, how come, with these particles, they couldn't get solid data on where they were, where they were going to go, or how they were going to get there? The best they could do was to get some odds on where something might show up or end up.

Of course, given enough particles and knowing the odds of any particular one ending up in any particular place, eventually a pattern will build up, and that pattern can be predicted with certainty. But that still left a very important set of questions without answers: what sort of particle is it that "knows" what's going on, changes behavior accordingly, and resists being known about so that you could never predict exactly what was going to happen with it?

It was assumed early on that this problem with measuring both position and momentum at the same time was due to the limitations of measurement technology. They thought that if the experiments could be figured out better, or if the instruments got more precise, eventually they could measure both speed and location accurately at the same time.

After working with different sets of complementary qualities and trying to explain it for a while, German scientist Werner Heisenberg came up with a completely new way of looking at the "fuzziness" problem. He said that it is an unalienable[40] characteristic of these particles that, to the degree an observer has knowledge about one complementary quality, that observer *cannot* know about the other complementary quality. And this applies to all the complementary qualities of a subatomic particle, of which there are many.

It's not a problem with or a failure of measurement; it's a fundamental quality of these particles, and by extension, of this universe.

---

40. Not capable of being taken away or denied. See Notes for citation.

This was Heisenberg's uncertainty principle, although I like the word *indeterminacy* better. It says that intrinsic to the basic nature of the fundamental building blocks of this universe is a certain minimum "unknowability" and randomness about them, a "not quite there-ness" that can never be resolved. They're nowhere near as solid as we'd like to think.

The number of particles and interactions that occur in our frame of reference is ginormous.[41] For example, there are roughly 2,560,000,000,000,000,000,000,000 atoms in a kilogram of uranium, the subatomic particles of which are interacting with other subatomic particles quadrillions of times per second.

Even though individual atomic-level events might each be random or fundamentally unknowable in some way, huge numbers of these events come together to form the predictable macro-level laws of physics and the reality we see every day. However, just because the reality we see all the time is apparently objective and predictable doesn't mean that the fundamental events they are built of are objective or predictable at all. It's similar to how the speed, direction, and position of any one molecule of water in a river might be totally random, while the motion of the river itself, made up of countless individual molecules, is quite predictable.

The uncertainty principle ignited a firestorm of controversy among scientists, with the contenders settling down into two camps. One side felt that all we were finding out with our inability to get accurate simultaneous measurements of the position and the momentum of these particles was that we needed to do a better job of measuring. The other camp felt that this fuzziness of measurement and the randomness of these particles showed something fundamental about this universe. They felt it showed

---

41. *Ginormous* is a very informal way of saying extraordinarily large or huge. Some things are more than huge—they're *ginormous*. See Notes for citation.

that, in some way, a certain minimum level of unknowability was an integral part of this universe. One famous comment was made objecting to this randomness, to the effect of "God does not play dice with the universe." Some felt that, if we could only find out more about these particles or how to better observe them, this objectionable randomness and unknowability about where they were or what they were would go away.

This controversy buzzed along without a real resolution until the mid-1960s, when an Irish scientist and mathematician named John Bell developed a mathematical proof that showed that, if one accepts logic as being an acceptable way of interacting with this universe (which it is, by the way), either these particles indeed have a certain inherent randomness and unknowability about them, or they could instantaneously influence each other over huge distances, or both.

To elaborate, somehow these particles either "knew" what was going on and were able to instantaneously respond to things that were happening huge distances away, or they were indeed somehow not quite "there" until they interacted with us, or both. Experiments have since been done that validate his proof. If you want to know more about this, look up "Bell's Theorem."

On a subatomic level, it's our decision about what to observe that partially determines what that thing we're observing becomes, where it is, and where it goes. At this fundamental level, the things that common sense tells us have a firm, separate, and objective existence are instead potentially one thing or maybe another, maybe here, maybe there, and it is only our interaction with them that causes them to become those things that we then see—and even then, we're not sure what we're going to get. *Until they interact with us in some way, the building blocks of this universe have no particular reality. And when they do interact, what they become is partially random and partially unknowable.*

These are data of universe-shaking importance. It means that things are most decidedly not as they seem. We don't live in a universe that exists independently of us. One way of looking at it is that we're each swimming in a sea of "becoming." Surrounding each of us is what you could think of as a "froth" consisting of subatomic particles, most of which are being observed or are interacting with those that are being observed. Almost all subatomic particles, because they're being observed or interacting with some that are observed, appear in a state of being. If interacting frequently, they continue in this state of being. It's difficult, but possible, for us to isolate a particle so that it's not being observed, can't be observed, and is interacting with nothing else that could be, or is being, observed. If this was the case, it could eventually drift back into being in a state of potential. However, even out in space, there are lots of ongoing interactions that keep the real universe humming along.

As solid as this universe may seem, it is the action of us looking, no matter how or when or where, that triggers the change of particles from potential to real. Looking triggers reality. No matter where we look, there it is.

This was hard to swallow . . . but there it was. The facts were in, and nobody could really dispute them. The experiments were done, and redone, and invariably came up the same way. The math was airtight and also borne out by experiment. These data were true.

Individual atomic-level events are inherently random and partially unknowable. The STUFF that makes up our universe does not come into existence without some participation by us. Even though each of these "becoming" events might be random or fundamentally unknowable in some way, huge numbers of these events come together to form the predictable macro-level laws of physics and our daily reality.

It also means that even though physical reality may seem to be stable and immutable, it's always a bit uncertain, and always open to some amount of influence from us. There's a certain amount of flexibility in the world. Maybe not a lot, but some.

So, all right, there's a certain amount of randomness and uncertainty built into every single particle of this universe. Do we have all the pieces we need to solve the puzzle yet? Nope, that's not enough. There's one more twentieth-century idea that we need.

CHAPTER 16

INCOMPLETENESS

We now see that subatomic particles can exist as potentials in multiple places or conditions (or no particular place or condition) at once, don't really become any particular thing until they interact with us in some way, and that some important part of their existence is always going to be unknowable. These ideas seem pretty outrageous. But we're not done yet. What's the final outrageous and unpredictable thing to come out of modern math or science that will help us solve the puzzle of reality? And how is knowing that going to help us sort it out?

One of the most skilled mathematicians and logicians of the twentieth century was Kurt Gödel. We touched on him and his work back in chapter 5, but we're going to get into his ideas in a bit more depth here. He was able to use mathematics and reason to prove some world-changing things. One of his biggest accomplishments was his "incompleteness theorem." Actually, there were two incompleteness theorems, but we're going to be mostly concerned with one of them.

In logic, a theorem is a proven statement based on certain basic data, and this one says that any system that uses logic and quantities of things can never be complete, in and of itself. This was a major challenge to the established ideas of mathematicians of that time who believed that math could be used to prove (or disprove) any statement in or about this universe. Because it established limits to the reach of mathematics (and shot holes in some cherished beliefs in the process), this theorem was controversial. However, because it could be verified by accepted premises, even though it was not a welcome idea, the validity of this theorem is now broadly accepted.

What does it mean when we say "using logic and quantities"? It means any system in which logic works in a consistent manner, and in which we can find quantities or amounts of things. As an example, a computer is such a system. So is a factory, a universe, an animal, or an atom, or almost any sort of mathematics using quantities—anything you can see or touch can be called a system, in addition to some things that are just ideas, like a system of philosophy or a mathematical system.

What about completeness or incompleteness? What does that mean? A system is considered to be incomplete if there are statements within the system that cannot be proven or disproven within that system.

Every system of this type has principles that are essential to the existence of that system. Those essential principles establish the system, and establish what can be true and what can happen in the system. These basic principles are usually described as being "self-evident" and are called axioms.

As an example of an axiom, let's use some basic math or logic. One axiom is that, for any two points, there is one straight line that passes through them. Another axiom is that things that are equal to the same thing are also equal to each another. These statements are obviously true, right? However, there's no way to *prove* these statements using existing math or reason. You just have to accept or agree that they're true. In math, these principles are usually pretty obvious. They're self-evident according to all our observations. Once you accept them, you can then build a whole system or field of mathematics using them or other similar principles.

Gödel showed that systems that use logic and quantities cannot prove their own consistency using their own rules. He showed that any such system must contain statements that are both true and unprovable, thereby rendering that system incomplete. This means they can't prove their own axioms. Why is this? If such a system were complete, it would be able to prove its own axioms. If such a system's axioms could be proven, then any statement within that system could be proven. The fact that there must be, within any such system, statements that are both true and unprovable means that no system that uses logic and quantities can prove its own axioms. Therefore, the fact that all systems that use logic and quantities must contain statements that are true but unprovable means that all such systems are built on axioms that they cannot prove.

The Implications of Incompleteness

The concept of incompleteness has some very interesting consequences, but these consequences might not be obvious. So, I'm going to sidetrack just a little here and go into Gödel and his ideas a bit.

Somewhere around 1960, Gödel wrote up a series of fourteen points that summarized the philosophical realities he had come to after a life devoted to science, math, logic, and reason. Three of these fourteen points were as follows:

- 10) Materialism is false.
- 13) There is a scientific (exact) philosophy and theology, which deals with concepts of the highest abstractness, and this is also most highly fruitful for science.
- 14) Religions are, for the most part, bad—but religion is not.

Why on earth would a man devoted to science and reason, whose peers were all physicists and mathematicians, after a life spent in careful thought, come up with these apparently "unscientific" ideas?

Gödel's theorem says (in essence) that a consistent mathematical system that can express[42] arithmetic cannot prove its own consistency. Furthermore, the Church-Turing thesis[43] says that a physical system can express elementary arithmetic just like a human can. As a result, the arithmetic of a physical Turing machine (a computer) is not provable within the system and is therefore subject to incompleteness. Taking this a step further, any physical system capable of measurement is capable of expressing elementary arithmetic. The physical universe (defined as the totality of known or supposed physical objects and phenomena throughout space) is capable of expressing elementary arithmetic, and so, like both mathematics itself and a Turing machine, must be incomplete.

Any system complex enough to deal logically with quantities, and which is based on fundamental principles or axioms, must necessarily be incomplete. The axioms it operates on must be accepted as unprovable in order for the rest of the system to be consistent (logical). The axioms of a system are the fundamental truths of that system on which the system is based, but being such, cannot be proven by the system itself. They must therefore, in some way, exist outside or beyond that system. This relationship is inevitable. Even if you expand the list of axioms enough to prove the initial axioms, all that happens is you end up with new unprovable axioms.

As a result, the axioms of a system can never be part of that system. They define the system. That's why a consistent system can never prove its own axioms, any more than a child can create its parents. One could say, "No

42. To show, manifest, or reveal. See Notes for citation.

43. Stated simply, the Church-Turing thesis holds that a computing machine is able to perform any calculation that any human can carry out. This thesis has not been proven, but it has always worked so far and is generally accepted. See Notes for citation.

system defines itself," or perhaps even (since the axioms have to come from somewhere), "No system creates itself."

The physical universe is a consistent system capable of expressing arithmetic. It is a system subject to this theorem. As a result, the origin principles of that system must somehow exist beyond or outside of it.

This explains some of what Gödel was talking about in the points mentioned above. Knowing that the system we call the physical universe is defined by axioms that exist beyond or outside of it, materialism, as we understand it, must be false (point #10). And if it is, there must be a place for an organized study of how these axioms came to be (point #13). And thus, there is a place for religion in our culture, not necessarily the sometimes mysterious and dogmatic religions that have gone before, but an enlightened study of what this thing we call a spirit really is and can do (point #14).

Here's another way of looking at point #10. Materialism assumes that the physical universe is a closed system, which means that nothing that is not part of that universe affects (or could affect) it. However, we know that any consistent system must have axioms that define it and are not part of that system. So, if this universe is consistent (workable natural laws, the validity of science, etc.), it must have axioms that exist beyond or outside of it. If there are no such axioms and the system is indeed complete and closed, it must therefore be inconsistent, and science and math would thus be invalid. As a result, one way or another, materialism has to be false.

So, we have a choice. We could reject science, math, reason, and the lot and live our lives based on individual truths and the emotions of the moment. Yet that experiment is being carried out right now in this society, and it doesn't seem to be playing out too well. Since science and math have worked pretty well so far for us, my inclination is to stick with them.

However, there are interesting and unforeseen consequences to this idea. If there are basic principles that establish a system but can't be proven from within that system, that means they could not possibly originate within that system. So, as discussed before, the axioms of any system that uses logic and quantities must have their origin outside that system.

This universe itself is based on certain basic principles. These principles are things like the existence of electromagnetism or the magnitudes of basic forces. We could ask, why does the phenomena we call electromagnetism exist, and why does it have exactly the force that it has? Can we use math to prove that it exists, or that it has to exist, or that it has to have the precise force it does? Actually, there's nothing to say that it *has* to exist or that it has to have exactly the force that it happens to have. There's no way to prove these things using math. Of course, if electromagnetism didn't exist, or was much stronger or much weaker than it happens to be, this universe would be very different and probably wouldn't exist as we know it at all.

The basic principles of this universe (or any universe or universes) must somehow have (or have had) their origin beyond or outside that universe, and these principles must be beyond anything that can be mathematically proven using data from that universe. When confining ourselves to just this universe, no data, no phenomena, nor any facts from within it can directly prove or disprove the basic principles that it runs on and with which it started (or, I suppose, where they came from).

If that's the case, then what? How do we proceed from here?

We can use observational data to show, generally speaking, that these principles exist and what their values are, but we can't prove that they *have* to exist, have to have those values, or where they came from or got started.

So, questions like "How did it all get started?" Why is it the way it is?" and "Was there a creator?" are all questions that lie beyond any direct proof (or disproof) using physical data. We can show or prove the effects of these principles, but we can't go beyond them. To use the language of cosmology, at the point of the singularity, the known laws of physics break down and no longer apply.

Recent thoughts from the physical sciences, such as the work from Hawking and Ellis (among others), all strongly imply that the material universe did actually have a starting point and started from nothing (a singularity) a long time ago. When we say nothing, in this case we mean nothing spatial, temporal, material, or physical.

Speaking of this universe getting started, the idea that it needed a push to get started is not a radical new idea. Aristotle (and many others) had the same thought, and all wrestled with it.

In any case, given all the data at hand, it looks very much like this universe didn't just sorta pop into existence one day by accident, nor did it create itself, nor has it always just "been." After all, no system can create itself. Somehow, it got started.

This makes sense when you look at how totally, fabulously, incredibly, absurdly, unimaginably unlikely it is for this universe just to exist or persist in any way. Any deviation whatsoever, no matter how inconceivably small (a trillionth of a trillionth of a trillionth of a bit), from the precise magnitude and interrelationships of the forces and components of this universe, and it would collapse instantly. It's beyond unbelievable that it's here at all. The fact that this universe is here and continues to exist makes winning the lottery twice a day, every day for twenty years straight, look easy and inevitable.

Personally, I don't know what started this universe, how it got started, or what was here "before" time existed. I don't know if this implies the existence of God, or gods. What I do know is that you can't prove *or disprove* the existence of God (or gods) using physical-universe data. I know you can't use physical-universe data to conclusively prove or disprove a creation event. Using nothing but physical-universe data to argue about these things is pointless. It's like using the data from a black-and-white photo to try to conclusively prove (or disprove) the existence of a certain color.

Hard data can only take you so far. Sciences in this universe are built on basic principles. Yet how can we ever know what the principles are for this universe if they arise outside the physical, and therefore can't be directly proven by physical data?

We can use science and technology to get enough facts, then ensure these facts are true enough to lead us to where we can use these facts and reason (as laid out in chapter 8) to derive[44] a good set of workable basic principles, then try them out and see which ones actually work the best. Science and technology, combined with reason and experience, can get us axioms. Then, with axioms in hand, the world lies open before us.

And with that, we have reached the end of the line in terms of startling or confounding[45] facts about this universe. Why did we bother with all these mind-bending ideas? After all, the current theories pretty much work fine. Granted, we might not totally understand what's happening, but we can already work with it to produce interesting (and sometimes even useful) technology. Why is a new theory even needed?

---

44. To reach or obtain by reasoning. See Notes for citation.

45. Throwing someone or something into confusion or disorder. See Notes for citation.

CHAPTER 17

THEORIES, EVIDENCE, AND THE EVOLUTION OF HUMANITY

The day science begins to study non-physical phenomena, it will make more progress in one decade than in all the previous centuries of its existence.

—Nikola Tesla

It is right that we should stand by and act on our principles; but not right to hold them in obstinate blindness, or retain them when proved to be erroneous.

—Michael Faraday

Those [scientists] who dislike entertaining contradictory thoughts are unlikely to enrich their science with new ideas.

—Max Planck

It's been nearly a century since the startling math and odd experimental results we discussed in previous chapters came to be (e.g., waves and particles; photons interfering with other photons that didn't exist; photons "knowing" if they were being observed; photons knowing about other photons huge distances away; indeterminacy, randomness when a particle appeared; superposition, etc.). It seems like we should have a good theory that explains how or why these things are occurring by now, right?

If not, what do we do? Should we bother trying to get a new theory, or be content with some adjustments to the old one?

Here's what I think: When any theory acquires too many loose ends, we need to put together a better one, even if we really like the old one. That's how human culture evolves.

A century or so ago, we thought we had a good handle on how everything in this universe worked. Then a bunch of strange and confusing data showed up and upset the apple cart of what was, then, "settled science." These revolutionary experimental results could be viewed as either a problem or an opportunity.

In the past, when experimental results cropped up that didn't make sense under established theory, that theory eventually got replaced or updated to embrace the new results, no matter how sensible the old theory seemed to be or how comfortable it was.

Take as an example the "flat-Earth" theory. Anybody could look out their window or go to the seashore and see that the earth was flat. The idea that Earth was a giant ball was nonsensical. This was a commonsense thing that everybody knew, and they knew it for many thousands of years. However, starting in the time of the ancient Greeks, some people started to do experiments that showed that maybe the earth was round. It took the coming of Columbus, Magellan, and others to really establish the idea that the earth was indeed round. Once people started to accept the idea of a round Earth, new doors opened to them, and a great age of exploration began. The "Earth is round" theory fit the facts, was simpler, and worked better than the "flat-Earth" theory, even though the earth sure does look flat from here. However, it did take quite a while for that idea to sink in and become accepted.

As another example, there was a similar change going from "the earth is the center of things" to "the sun is the center of things" theory. The Greek astronomer Ptolemy said that we were motionless and that the sun, the moon, and the stars all moved around us in perfectly circular orbits. Anybody can look up and see the sun, the moon, and the stars all moving overhead, circling around us. Also, anyone can look at themselves and see that they are motionless while all these celestial objects go soaring around. It's just common sense. Of course, there were a few bugs in the system, such as, how come some of these heavenly bodies looked like they stopped and traveled backwards every so often? Those observations were addressed with new wrinkles tacked onto the old theory. In the case of the planets and some other objects, the idea of circles within circles was added to the theory to address the issue of them going backwards from time to time. It added complexity and clumsiness to the system, but it seemed to account for the motion. Here's a diagram of how it looked:

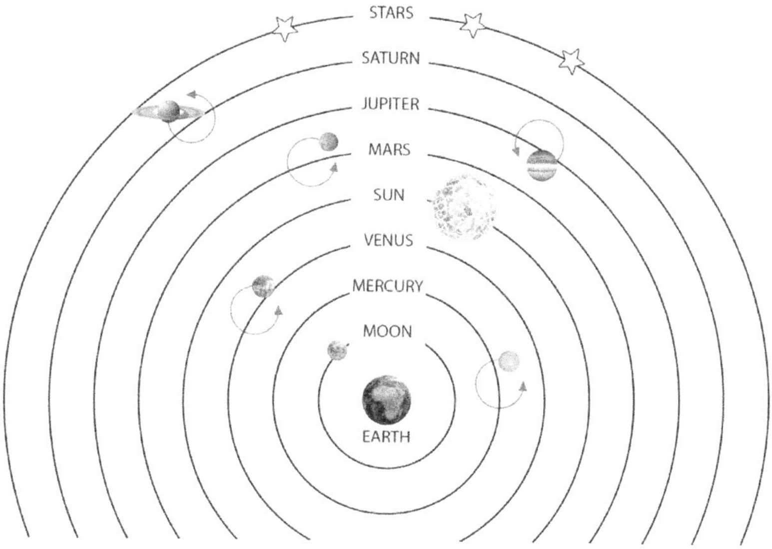

Figure 9: Ptolemaic model of the solar system.
Source: William Truax.

This theory got more developed over time and eventually became quite complicated. This was because lots of "wrinkles" had to be added to it in an attempt to address all the little discrepancies that kept showing up.

More than a millennium after this theory came into common acceptance, the new "sun-centric" model emerged, which said that Earth is one of the planets orbiting the sun, instead of vice versa, and that all the planetary orbits were ellipses,[46] not circles. The transition was rocky, but over time it became undeniable that the new theory fit the facts better, was simpler, and worked better than the old one. Eventually, the new theory was accepted and the old one consigned to the "nice try" bin.

There were a couple of leaps in thinking needed to make the switch from the old Earth-centric theory to the newer sun-centric theory: first, the idea that what appears to be happening isn't what's actually happening; and second, the idea that there's this invisible force or effect called gravity, which makes it look like all things are attracted to all other things. As improbable as they may have seemed at the time, understanding and accepting those phenomena was necessary to a better understanding of the universe.

Both the "flat-Earth" and "Earth-centric" models are examples of cases where an older, commonsense way of looking at the world, while useful, eventually was undone by the fact that it had to rely on "add-ons" or kludges[47] to explain newly observed facts. Once these new observations were accepted as factual, all that remained was for a new and better theory

---

46. A closed . . . curve shaped like an oval. See Notes for citation.

47. A software or hardware configuration that, while inelegant, inefficient, clumsy, or patched together, succeeds in solving a specific problem or performing a particular task. See Notes for citation.

to emerge that didn't rely on sticky patches, zip ties, and duct tape. Once that new theory emerged, the doors to progress opened wide.

Current Theories

The odd phenomena that popped up in atomic physics were a mystery when they first appeared. And in many important ways, they remain a mystery. Our current theories don't do a good job of simply, clearly, and completely explaining why these things happen. Make no mistake; we have lots of theories, and all of these theories have some workability. A number of different theories have been proposed, and while all of them have strengths, all of them have been lacking in some way or another, and almost all of them are very complex. The aroma of add-ons and kludges hangs about.

The most widely accepted theory is called the "Copenhagen interpretation" of atomic physics. The first third of the twentieth century was the time when the experimental evidence coming out of atomic physics was fresh and startling, and there was a lot of trial and error that went on as physicists who were firmly grounded in the "old" physics tried to deal with these crazy new facts. The Copenhagen interpretation was worked out by a group of atomic physicists around the world in the early twentieth century. One of their leaders, Niels Bohr, was based in Copenhagen, and many of the ideas that led to this interpretation came out of an institute he ran in Copenhagen.

In very simple form, the Copenhagen interpretation says that, at a subatomic level, matter and energy each have two corresponding but completely different sides. If you do an experiment that looks for particle phenomena, you get particle phenomena. If you do an experiment that looks for wave phenomena, you get wave phenomena. The theory and math for the particle side of things works out beautifully and correctly predicts all sorts of particle-related things. However, if you do the math and the theory for the wave side of things, it also works out and predicts all sorts of wave-related things. Both models are completely different and don't interact with each other. They are two completely separate and different ways of dealing with atomic-level phenomena, but both work.

The Copenhagen interpretation doesn't attempt to reconcile these two very different sides of atomic reality. It just says, "They both are," and uses them for different things. This allows scientists to use whichever model and math suits what they're doing at that time. In a way, that's good, because these models and math have been instrumental in much of the tech revolution of the last fifty years. These models are used in computer-chip design and GPS, and those two things alone are responsible for as much as 30 percent of the world's economy.

Even though your cell phone works, this model doesn't provide a clear and comprehensive overview of how the universe works. For that purpose, it's about as useful as stating, "We don't really know if you're standing on the floor or the ceiling; in fact, we think that what you're standing on could be either the floor or ceiling or both, depending. However, we do know the floor model works great if you take the viewpoint that you're standing on the floor, and that the ceiling model works great if you take the viewpoint that you're standing on the ceiling." While this may be well and fine as an idea, it's not much help to someone who's trying to find the door.

So, the math and technology work. But . . . this theory doesn't provide a clear and comprehensive overview of what's actually going on. It just acknowledges that we're dealing with two different things at the same time and goes from there, leaving a mystery in its wake.

You could say that, since it doesn't even try to explain the entire universe but just parts of it, this theory must be lacking. Fair enough. However, all the other theories put forth to explain these odd occurrences have had problems of their own. With all current theories, we have ideas that help us deal with the paradoxes of what's happening without really explaining how or why it's happening or that can't explain it without encountering some other serious issues. They violate one or more of our rules of logic. As a result, there is currently no simple and comprehensive explanation for all these phenomena.

For example, the "many worlds" interpretation holds (in essence) that the fuzziness that occurs in our observations is a symptom of two (or more) things both happening at once, at which point the universe branches off into multiple possible futures. Since the universe as we know it is made up of a fantastically huge number of particles, and since these particles interact with others trillions of times per second, that means there's a virtually infinite number of alternate universes (quadzillions!), a number that is getting "more" infinite with each passing second. None of these alternate universes, nor any trace of them, has ever been, or ever will be, detected.

Another current model, called string theory, answers questions other theories leave open by stating that all particles and forces come about as a result of one-dimensional objects (strings) folding and vibrating in different ways. But it requires that we postulate ten dimensions of space and time, and we have yet to see any trace of six of these dimensions. The math is cool, but the theory lacks something.

Yet another model holds that none of these phenomena of reality actually exist at all, except as information that can be manipulated and predicted, while another holds that logic is actually very different from what we normally think of as being "logical."

Moreover, there are others—lots of others—with new ones coming all the time. Interestingly, almost all of them are based on the same fundamental concepts of materialism, but with some developments or "wrinkles" added on. And all of them have at least one element that is complicated, incomplete, or difficult to reconcile with reality.

I think we can develop a better theory of what's going on, but doing so will also require some leaps in our thinking.

Ready?

CHAPTER 18

A NEW MODEL OF REALITY

In all my research I have never come across matter.
To me the term matter implies a bundle of energy
which is given form by an intelligent spirit.
—Max Planck

In the great drama of existence we are
audience and actors at the same time.
—Niels Bohr

So, what is reality?

We've covered the older materialistic model of classical physics, which, while popular, doesn't actually fit the facts too well. We've also looked at newer models that attempt to explain quantum phenomena, but these also have issues. Knowing what we know about using the shortcomings of an existing model as an indication that it's time to evolve a new one, the time has now come to see if we can do just that.

Reasoning Beyond the Physical

Let's recap some of the principles of reason from chapter 8. Science, as we have practiced it, has been very good for (and to) humanity. It's given us lots of truths, lots of technology, and lots of prosperity. Yet, as a culture, we may have made a mistake. As a reaction against the earlier failings of religion and philosophy, we decided to use physical data exclusively to validate what we would accept as truth. While that was certainly an advance over prior educated guesses (or opinions), it was itself also somewhat limiting. While STUFF is important, there's more to life than just STUFF. Part of the answers to the really big questions lies beyond the realm of the strictly physical.

Restricting ourselves to physical data restricts us to physical truths only. We can know all about STUFF, but no more, and what we're looking for goes beyond STUFF. But if we can't validate our thinking by referring back to STUFF, how are we to know that we're on the right path—the path to answers to the big questions that have some truth, and that actually work for us?

The true lessons of science are not so much what to know or what to think or how to make things, but how to know and how to think. STUFF is nice, but the real prize we're after as a race is more truth, bigger truths, and more basic truths. We can use the lessons of science to reach for those things. We can use the lessons learned from science to know how to think, how to evaluate data, and how to recognize truth. We can take on the realm that goes beyond raw physical data, and take it on with some certainty that we can now find our way where before we had always gotten lost.

There are a few techniques we can use to validate our thinking when we venture to stick a toe into the transcendent. Probably the most important is the *lex parsimoniae* (the law of simplicity), which was discussed in chapter 8 and can also be known as Occam's razor. This is a principle that says the best explanation is the one that requires you to make the fewest possible assumptions about what's involved. Another way to say this is that the correct explanation or solution is usually the simplest. If you're looking at a bottle of milk and wondering how it got on the countertop, thinking your spouse did it is much more likely than thinking that it walked over there by itself. Self-propelled milk bottles involve far more assumptions than does spousal action.

The law of simplicity is not just a coincidence or a convenience. It's a statement based on how this universe is put together and how it works. The law of simplicity should be applied to the unusual experimental data from atomic physics and the developments in math to see if there could be an explanation for these strange phenomena that is simpler, more fundamental, and more comprehensive than any that we, as a civilization, have had so far. Basically, looking at all the new data and knowledge we have, what's the *shortest* line that connects all those points?

Another technique is the principle of causal adequacy. This is a principle that says that the cause you're looking at to explain some phenomena must actually be able to cause the phenomena in question. For example, looking at an operational orbital rocket and thinking that it was built by a bunch of eight-year-old children will probably violate the principle of causal adequacy. No way could a bunch of children that age build a rocket capable of achieving orbit.

Another technique is prediction. If you have the right explanation, or somewhere near the right explanation, it should work to produce and predict results. Another way of saying this is, try it out and see if it works. Does it produce the expected result? Does it correctly predict other phenomena?

The final technique is the application of the principle of inference[48] to the best explanation. When you're looking at multiple possible explanations for something, and you're precluded from being able to prove (or disprove) them with certainty, you pick the explanation that does the best

48. Infer: to derive by reasoning; conclude or judge from premises or evidence. See Notes for citation.

job of actually explaining what's going on. Does it answer more questions for you than other potential explanations?

So, can we use reason to move beyond what can be verified with physical data? I think so. In summary:

- Is it the simplest explanation with the fewest moving parts or unproven assumptions?
- Is it capable of actually producing the effect in question?
- Does it predict things that are later found to be true?
- Does it work?
- Does it answer questions that otherwise would go unanswered?

Knowing that we exist and accepting this universe as real, we can use these methods to nail down some workable truths. However, this method will work only as well as we do. It depends on us for accurate observations, honest evaluations, and a devotion to truth. If we do those things, we can resolve many of the mysteries of life and this universe and use this data to make our lives better.

Data

All the knowledge we had about physical phenomena before the twentieth century had been built up using substances composed of huge masses of particles. We did not really get a good look at the subatomic realm until early in the twentieth century. It was only once scientists were able to do experiments looking at, and working with, one single particle at a time that they noticed some new and unsuspected properties of the subatomic world. Here are some things that have been established:

- Our inability to know certain sets of subatomic-level data simultaneously is not a failure of measurement but is instead a basic characteristic of this universe.

- What we call subatomic particles don't actually exist as real and recognizable objects (electrons, photons, etc.) unless and until they are observed, observable, or interact with some other particle that is already extant.[49] For our purposes here, we'll call either an observation or interaction event an "interaction."

- Until that interaction, those things we call particles exist in a state called superposition. In this state, the particles have no particular location or characteristics. Using location as an example, a particle in superposition could be said to exist everywhere (or anywhere) within a given volume of space. Prior to interaction, any particle has a large number of potential locations and characteristics. It doesn't exist as a single recognizable "thing."

- We can't predict where a subatomic particle is going to show up upon interaction, and what its characteristics will be when it does. What and where it is (or becomes) is always partially random and depends in part on what we do and what we're looking for.

- Interaction causes the particle to lose most of its potentials and actualize as mostly a single identifiable thing with particular properties, such as a location. We talk about this as the particle having "become" a particular thing in a particular place. However, even in this condition, the particle has not lost all of its potential properties. It is not entirely in one place. It has not entirely

49. Still in existence; surviving. See Notes for citation.

become a single thing. It still retains some very small quantity of other potentials.

- After the interaction takes place, the particle gradually reassumes its original potentials, becoming less an identifiable thing and more a set of potentials. This continues until the next interaction.

- We usually never see one single particle all by itself (which is why we never noticed any of this before)—we see lots of them, all interacting. Subatomic particles generally interact with other subatomic particles countless times per second.

This incredibly rapid and continuing interaction generally makes these subatomic potentials act and look like identifiable and continuing real objects.

Although the basic particles and motions that make up our universe are irreducibly random in some way, normal-scale events that are very predictable come about by combining large numbers of random events.

Pairs of particles can be interlinked such that, when we observe one of a pair of these interlinked particles, thereby causing it to actualize in some condition (such as spinning "clockwise"), the other member of the pair actualizes in the opposite condition (spinning "counterclockwise"[50]) without any time involved, regardless of distance. This interlinking is called entanglement or "action at a distance."

These data show that the idea of a completely objective physical reality must be incorrect or incomplete. Otherwise, particles would not be indeterminate until we interacted with them in some way, or it would be

50. Subatomic particles don't actually "spin" in any sort of classical sense, and they certainly don't spin clockwise or counterclockwise. However, they do possess an identifiable quantum property that scientists call *spin*.—William Truax.

impossible for particles to instantaneously influence each other over huge distances. Experimental data and logic have clearly validated indeterminacy and the instantaneous influence particles can exert over each other across huge distances. Therefore, it can't be that the universe exists "out there," completely separate from us. This has to be the case because our actions have an effect on "what's there." As Gödel said, "Materialism is false."

The Beginnings of an Explanation

How can we start to explain these facts in a way that is the simplest and most respectful of the law of simplicity, as well as other techniques of reason?

- What we call a subatomic particle is fundamentally a tendency to be or exist, which is characterized by having many potential properties, all at the same time. This particle can actualize at one point or another with (mostly) one particular set of properties. It's not an eternal little marble that just "is."

- The actualization of the particle is brought about by an interaction with an observer or some other particle. Absent interaction, the particle is nothing more than a potential. Even if it's an interaction with another particle, that interaction is ultimately an interaction with an observer, or something that is being, can be, will be, or has been observed by an observer. Down at the beginning of the chain, an observer has to cause some particle to actualize, which can then start a chain of interactions and actualizations. No particle actualizes on its own.

- Observation requires an observer who is separate or different from what is being observed.

- The second incompleteness theorem shows that this universe cannot create or have created itself. That which creates this universe and its operating principles must, in some way, transcend[51] it. As a result, that which creates this universe and its operating principles must be beyond it, different from it, outside it, or separate from it in some way. The creative principle must necessarily, then, be nonphysical.

- Subatomic particles interact with each other countless times per second. This rapid and continuing interaction makes these subatomic potentials tend to settle down into stable arrangements, which can be viewed, and their reality agreed upon, by multiple observers. Some of these arrangements are more stable than others. The less stable arrangements "decay."

Taking all these things into account, we can postulate[52] as follows:

- Particles in this universe are being created "in real time" by a separate, *nonphysical* creative force that, while acting on (and in) this universe, also transcends it.

- Applying Descartes's idea of "I think, therefore I am" to this data, we could modify it slightly and state it as, "I observe, therefore I am." If that's true, and our observation is, in some way, what causes subatomic particles to actualize into what they are, then we could also say, "I look, therefore it is." *Aspicio, ergo est.*

---

51. To exist above and independent of (material experience or the universe). See Notes for citation.

52. To claim or assume the existence or truth of, especially as a basis for reasoning or arguing. See Notes for citation.

- Particles in this universe are indeterminate because the only way for a created thing to not be eternally predetermined is to be created so that it somehow runs independently of its creator(s). Making it somewhat random or out of control accomplishes that. As participants in the grand system of creation, we can be part of the ongoing creation of particles while still keeping them just a bit beyond our grasp. That gives us the ability to participate in the creation of this universe while still being able to be "in" this universe and subject to its rules. That allows us to have meaningful free will and a shared reality.

- Being partially unknowable is what enables a particle to exist continuously in time. By being unpredictable, moment to moment, a particle has to be rendered new and newly in a series of succeeding moments.

The New Model of Reality

So here it is, in all its glory, the currently shortest line between these data points, the new model of reality:[53]

- What we call a subatomic particle is fundamentally a tendency to be or exist—not an object. This tendency to be can (and does) come into recognizable being—it can actualize.

- The actualization of the particle is brought about by an observation or an interaction with some other particle, which is ultimately an interaction with an observer, or something which is being, can be, will be, or has been observed by an observer. An

53. This model includes several points mentioned earlier, but they're part of the complete model, so they're included again here.—William Truax.

observation is required to start the chain of actualizations into motion. Particles do not actualize on their own.

- We can't really predict where a subatomic particle is going to show up upon interaction, and what its characteristics will be when it does. What it is (or becomes) is partially random and depends to some degree on us: what we do and what we're looking for.

- Interaction causes the particle to lose most of its potentials and actualize as mostly a single thing with particular properties, such as a particular location. We talk about this as the particle being or having become a particular thing. However, even in this condition, the particle has not lost all of its potential properties. It is not entirely in one place. It has not entirely become a single thing. It still retains some small quantity of other potentials.

- After the interaction takes place, the particle gradually reassumes its original potentials, becoming less a single identifiable thing and more a set of potentials. This continues until the next interaction.

- This condition of a particle appearing to exist in the separate states of "potential" and "actual" can lead to confusion, since we call both the potentialized state and the actualized state by the same word—particle. Most people associate the word *particle* with a thing, not a set of potentials. As such, for purposes of this model, we will call the potential state *proto-particle*, leaving the word *particle* to describe only its actualized state.

- Proto-particles interact with each other countless times per second. This rapid and continuing interaction makes these subatomic potentials settle down into stable arrangements as particles, which can be viewed and their reality agreed upon by multiple observers. Some of these arrangements are more stable than others. The less stable arrangements "decay."

- Any universe that uses logic and quantities, including this one, is necessarily incomplete. Its origin and founding principles must transcend it or exist, in some way, outside it. For a physical universe like this one, transcending it or being "outside" it means existing while being nonphysical. While this may sound odd at first, lots of nonphysical things exist, such as ideas.

- Particles in this universe are, in some way, being created "in real time" by a separate, nonphysical creative principle or force.

- We, each one of us, are all individual manifestations of this nonphysical creative principle or force. In prior philosophy, this concept has been called spirit, soul, or *élan vital.* The ultimate actuality of each of us transcends the physical.

- We each, via observation, play some role in the creation of the subatomic particles that comprise this universe.

- Any particle in this universe is ultimately made up of "nothing" aside from potential (motion) and an element of unknowability.

- Being partially unknowable is what enables a particle to exist continuously in time. The particles themselves have no inherent persistence. By being unpredictable, moment to moment, a particle is rendered new in each succeeding moment of interaction, thereby creating a series of moments—time.

- Some types of particles have a greater tendency toward interaction than others.

- Space implies an "out there-" or "over there-ness." For the concept of "over there" to be meaningful, there has to be an observer "here" and some thing "over there" to observe. In this universe, the things that are "over there" are either particles or composed of particles. Space is created by the observation of particles "over there" that have been made different from the observer/creator.

- We don't have to know the specifics of exactly who or what or how all this creation is getting done (God, gods, me, you and me, some combination of these, etc.) in order to use this data to resolve many issues of concern to humanity. All we must know is that creation is being done by that separate, nonphysical creative principle or force that will be called *essentia*[54] or LIFE (small caps), and that we are each individual manifestations of this *essentia* or LIFE. LIFE is something like beauty, in that it exists and acts in the present but is separate from, and senior to, this universe.

- LIFE can know, create, decide, and impel, and can consider that it is located in this universe. Time and space, energy and matter, are all products of LIFE. Thus, LIFE is always senior to STUFF. Time and space, energy and matter, will be called *materia*,[55] or STUFF.

- All activity and existence in this universe are the product of the interaction of two opposed (or polar[56]) principles or qualities, which grow out of the inherent differences between LIFE and STUFF. The most fundamental of these is the division between "that which creates" and "that which is created."

---

54. Latin: essence, being, essential thing. See Notes for citation.

55. Latin: material, stuff. See Notes for citation.

56. Opposite in character or action. See Notes for citation.

- That division provides the basis for all other polar qualities, which spring from the basic nature of "that which creates" as well as "that which is created." For example: alive vs. not alive; order vs. disorder; awareness vs. unawareness; enthusiasm vs. apathy; plus many others.

- While these polar qualities do not themselves exist in pure form in this universe, they combine to form the qualities that do exist in this universe, and the existing combinations of those qualities always exist somewhere along a continuum[57] between the two extremes.

Consequences

This model is, in some respects, a very different way of looking at reality from what you may have been accustomed to. However, it seems to be the best answer we currently have for all the questions and paradoxes presented to us by modern life, science, and math. Following through with it gives rise to some very interesting and important realizations and/or consequences:

- We each participate in the ongoing creation of this universe. It's nothing without us. As such, we have some responsibility for not only ourselves but for the wider world around us.

- The purpose of LIFE is more LIFE. This does not necessarily mean just more life forms. It means more of those things that characterize LIFE: things like beauty, order, happiness, enthusiasm, and such. These purposes as and for LIFE take priority over any other purposes in this universe. Handling STUFF can certainly have

57. A continuous series or whole, no part of which is perceptibly different from the adjacent parts. See Notes for citation.

value, but these actions should always be taken within the framework of a higher purpose—a LIFE purpose.

- Any activity occurring in this universe has to be understood fundamentally in terms of the purposes and actions of LIFE in order to handle it with ultimate success.

- No effect occurs in this universe that isn't caused.

- The way things get better is only if we make them better. Our sciences should be arranged so that they are based on achieving those things that are of the most value to us who participate in the creation of the physical, rather than what would be important to creatures who were nothing but the physical.

- The only limits to our potential are those that we impose on ourselves by our beliefs, fears, or decisions, or that are imposed on us by our lack of knowledge.

- Every single thing in this universe has an element of disorder or out-of-controlness embedded within it. It has to be that way in order to exist as it does. This gives the universe a tendency toward contrariness (Murphy's law[58]) and deterioration. It also means that, starting from any given point or condition, if you move in the direction of more LIFE, you will encounter barriers that will need to be overcome.

- *None of this changes the reality of this universe or the workability of its laws and technologies.* The universe is just as real as ever, all its

58. An observation: anything that can go wrong will go wrong. See Notes for citation.

laws are still operational, and our lives are still transpiring within its bounds.

- However, new laws, theories, and techniques can now come into view. As a result, the answers to life's mysteries can now be more accessible to us. Using them, life can be understood and better lived.

Summary

We are not just, or even primarily, intelligent or evolved animals. We are spirits, with minds, who inhabit evolved animal bodies. As spirits, we participate in the creation of reality. This creation is done in a manner so as to result in a real universe of time, space, energy, and matter. There are many qualities inherent in "that which creates," as there are in "that which is created." These qualities are always the opposite of each other. This universe exists as an interplay between the qualities of "that which creates" and "that which is created." Any existing quality (such as the relative orderliness) of any individual thing that exists in this universe is a combination of these two extremes (in this case, some order and some disorder). Any activity occurring in this universe has to be understood fundamentally in terms of the purposes and actions of LIFE in order to handle it with ultimate success. None of this changes the reality of this universe or the workability of its laws and technologies. We still live here and have to deal with reality, but we now can develop tools that allow us to move toward mastering that relationship.

There you have it, a new model of reality, which spells out our place in this world. But how do we know this is not just some fevered dream?

CHAPTER 19

VALIDITY

I regard consciousness as fundamental. I regard matter as derivative from consciousness. We cannot get behind consciousness. Everything that we talk about, everything that we regard as existing, postulates consciousness.

—Max Planck

The brain is a computing machine connected with a spirit.

—Kurt Gödel

Nice theory.[59] . . . Why should we grant it any credence?

Well, it's pretty clear that you can only go so far with strictly physical data. We've seen that there must be something more than, or something which reaches beyond, the strictly physical. As a result, to make better sense of our universe and our lives, we must be able to dip a toe into the waters of the metaphysical (by the way, that word actually only means "beyond the physical"). However, whenever humanity has done this in the past, there's tended to be problems. Lots of dogma, opinions, and unverified assertions that ended up leading to bad places.

Obviously, we don't want to do that now. But how can we touch the metaphysical without bedding down with corrupt data? How can we

59. I know . . . it's actually a hypothesis.—William Truax.

develop a theory that goes beyond strictly physical data without flying off into some ridiculous unreality?

We have lots of tools, as given in chapters 8 and 18, that allow us to separate more suitable candidates from less suitable candidates. We can see which candidates cover all the bases without leaving things unexplained, which candidates could reasonably cause all the effects in question without too many mental gymnastics, which candidates require the fewest "special cases," and which candidates can answer old questions and predict new phenomena. Having done all that, we can then see which candidate is the simplest, explaining the most phenomena with the fewest assumptions.

One problem that pops up is, "How do I do a physical test on something that is inherently nonphysical?" Fortunately, some things can be tested indirectly. That test is, "Does it work?" We can see which candidate works the best when put to the test in the real world.

We know that the experimental data coming out of atomic physics defied explanation back in the twentieth century, and much of it continues to defy explanation today. We have some theories that cover some (or many) of the bases, but there are still huge gaps in our understanding. Some work needs to be done.

For example, how can we explain wave-particle duality, the idea that these subatomic things are waves and particles at the same time? How about entanglement? How about superposition, indeterminacy, and what scientists call "the collapse of the wave function" (that's when a particle stops being a potential and starts acting like a thing). How about incompleteness? How do we explain these things?

I've picked a few of the best current candidates from the world of science and put them up against the new model from chapter 18 so you can see how they all stack up. I've put it in table form to make it easier to digest.

The theories under inspection were touched on in chapters 10 and 18. They are materialism, the Copenhagen interpretation, string theory, the many worlds theory, and the "new" theory, which I'll call *essentia* theory.

The following criteria will be used in comparing them:

- Comprehensive? Does the theory cover all the bases? Does it explain everything we're seeing?
- Causally adequate? Can the theory produce the effects we're seeing without too many mental gymnastics, far-fetched explanations, or special cases?
- Testable? Can the theory be tested, directly or indirectly? Or does it depend on making some sort of assumption that we'll never be able to verify, or depend on some postulated phenomena that we'll never be able to see?
- Predictive? Does the theory predict things that are later found to exist?
- Simplicity? Which theory explains the most phenomena with the fewest new assumptions?
- Workable? If put to the test, does it work? If so, how well? Lots of things sound good, but sounding good is not good enough. They have to work well too.

Theory Name	Compre-hensive?	Causally adequate?	Testable?	Predictive?	Simple?	Workable?
Materialism	No	No	Yes	Not with sub-atomics	Yes	Limited
Copenhagen	No	Yes and No	Yes	Yes	Yes	Limited
Many worlds	Yes	Yes	No	Unknown	No	Unknown
String theory	Yes	Yes	No	Yes and No	No	Unknown
Essentia theory	Yes	Yes	Yes, indirectly	Yes	Yes	Yes

Materialism and the Copenhagen interpretation have a lot going for them. They're simple and straightforward, testable, and have decent workability—up to a point. However, they flunk the biggest test of all (for our purposes). They cannot even begin to explain why what's going on in the subatomic realm is what's going on. As far as they're concerned, the subatomic world is all a mystery. Maybe a mystery that is predictable in some ways, but still a mystery. So, they're out.

That leaves us with the many worlds theory, string theory, and *essentia* theory. While the many worlds and string theories do offer an explanation of what's going on with the subatomics, they rely on phenomena that nobody ever has seen (or will ever see) or assumptions that can never be verified. They're not testable. Plus, sometimes they come up with predictions and answers to other questions that don't gibe with reality. To the degree they predict anything in this universe, their predictions are sometimes right and sometimes wildly off. As a result, we can't tell if they would actually work or if they're just creations of our imaginations.

More fundamentally, the many worlds and string theories are essentially add-ons to materialism. They start by accepting at least one of the basic postulates of materialism, and then add on special cases or conditions that attempt to explain what's going on at a subatomic level. It's a bit like the add-ons and adaptations made centuries ago to the Ptolemaic model of the solar system, which were attempts to keep that system relevant by postulating variable motion, circles within circles, and other additions to the basic model, thereby hoping to explain observable data that the basic model itself could not explain.

As a result, the many worlds and string theories violate the law of simplicity. Instead of coming up with a straightforward new way of looking at things, they're both tacking "explanatory" elements onto an existing theory, which has run across some things it can't explain. The many worlds theory requires that we postulate an increasingly infinite number of invisible alternate universes into existence, such that every possible outcome of any subatomic interaction is represented by a separate universe. String theory requires that we postulate one-dimensional strings existing in ten dimensions into existence. Not only will we never see any of these alternate universes, one-dimensional strings, or the ten dimensions they exist in, we can't even test for them, directly or indirectly.

Perhaps most importantly, all of these theories attempt to avoid the question of ultimate causation. Here's a universe—was it caused? Why is it here? They all have as a basic assumption the concept that the universe just "is," then go from there.

That leaves *essentia* theory. It offers a cogent explanation for what's going on in the subatomic realm. It satisfies causal adequacy, in that it's capable of causing the effects we see. It's testable, albeit indirectly. While we can't create a direct physical test for something that is inherently nonphysical, we can test for the effects of such a thing, and these effects do seem to exist. It predicts things that, when looked for, can be seen to exist. It obeys the law of simplicity, in that it only requires one single new assumption—the existence of individual manifestations of a nonphysical Life Force (*essentia* or spirit), which can decide, know, create, and impel, and considers that it is located in this universe. In doing so, it explains many more mysteries and answers many more previously unanswerable questions (more on that below) than any other theory. It even partially addresses the question of origins. And, because of that, it's workable, and works better than any other theory.

So, if we take the experimental data from science, combine it with advances in math, and evaluate the whole mixture using the rules of logic, we come up with one answer that is notably better than others. From the point of view of reason and logic, the most viable theory for observed subatomic phenomena, as well as all existing larger-scale phenomena, is *essentia* theory. It's the best answer we have to the mysteries and paradoxes of the subatomic realm, as well as being the best answer to some of the mysteries and unanswered questions of the human sciences.

Other Questions Answered

Chapter 18 addressed a lot of previously unanswered questions about the subatomic realm and showed that *essentia* theory provides answers for them. It also hinted at the fact that this theory might provide answers for other fields of human endeavor as well. Like what?

- **The mind-body problem.** One of the reasons Descartes's philosophy fell on hard times is that he couldn't persuasively explain how something nonphysical could influence the physical. I think we've addressed that question; but even if you believe we haven't completely answered it, we've certainly made distinct progress toward closing the circle.

- **What is reality?** I don't have enough data to say for sure, but it looks like it's the combined and compounded end result of mutual creation. You participate in the creation of particles, I participate in the creation of particles, you create a world according to certain rules, I create a world according to the same rules, and so on, until the end result of all this combined and (mostly) harmonized creation is the reality we see about us. Because we all participate in this creation, the world belongs, in large measure, to us. We each make decisions about how we think things are or should be, and those decisions influence reality and create it as we think it should be, for better or worse. Some of us are better at this than others.

- **Why do some people seem to have all the luck?** As noted above, some of us seem to be better at making decisions and influencing reality than others, although increasing this skill does seem to be possible for anyone.

- **Is there actually such a thing as a human spirit?** That's the answer that makes the most sense of the existing evidence and provides the best explanations for the mysteries of life and this universe. Actually, a better statement of it would be that there's a creative force and potential that exists separate from the physical universe, but that is "in" this universe, and we call that creative force or potential LIFE. LIFE is that which animates all living things and, through its own creation, participates in bringing this universe into existence. Each human spirit is an individual manifestation of LIFE. You can certainly come up with other answers for the way the world works, but as scientists sometimes say, these other explanations need a few miracles in order to hold water. Any system that needs miracles to hang together is probably in need of a revision.

- **Which side of life is more important—the physical side or the spiritual side?** The spiritual side. That's because it's the spiritual side that brings the physical into being. The physical universe is an artifact[60] of the spiritual side of life.

- **Why is there a placebo effect? How does it work?** We directly influence the reality that occurs around us with our decisions, and those decisions are driven by our hopes, our intentions, and our fears. To the degree we have that influence, it is possible, in some cases, for us to create ourselves well again.

- **Why does entropy[61] exist?** STUFF is created with an unalienable element of disorder. It wouldn't exist as we know it without having that embedded fundamental disorder. Without LIFE bringing order to it, STUFF will always fall back toward its native state of disorder.

- **What is consciousness?** While there may be some sort of overreaching or cosmic consciousness, the most important thing to know is that we are each individual manifestations of a nonphysical creative force called LIFE or *essentia*, and that this force is, itself, natively conscious. It is, it perceives, it decides, it's capable of knowing, and it thinks. Consciousness is not your brain—it's you. You are aware. You can and do use a brain as part of the control system for that composite entity called a human being. Why composite? It's essentially an intangible "me" working with a partially tangible mind, which acts as an operating system for a fully tangible body.

60. Any object made by human beings. See Notes for citation.

61. (i) A state of disorder, or a tendency toward such a state; chaos. (ii) A doctrine of inevitable social decline and degeneration. (There are other, more technical definitions that also apply.) See Notes for citation.

- **What is the relationship between science and religion?** Science is a set of known things, techniques for acquiring knowledge, the organization of that knowledge into sensible and workable bodies of knowledge, and the discipline to do it all correctly. Religion is that body of knowledge that involves the spirit, especially the human spirit, its identity, its capability, and its purposes. It can include a God or gods, but that's not required (e.g., Buddhism). Religion is *not* just "belief in something unprovable or some dogma." All of the philosophical discipline of science applies to religion. Religion can (and should) be logically organized, disciplined, and effective. Faith, within this context, could be defined as "knowledge of nonphysical truth."

- **Why insist on using reason and logic to deal with life? Aren't other approaches just as valid, if not more so?** I use reason because that's how this universe is built. Around here, 1+1=2. Of course, you could always go for an approach that uses something other than reason, but you're going to be like a medicine man trying to fix a broken automobile. You may have the best intentions and really make a good show of it, but without the operator's manual, you're not going to have much success. Reason is what works in this universe.

- **If a tree falls in the forest and there's nobody around to hear it, does it make a noise?** Provisionally, yes. Assuming there are some LIFE entities around somewhere, reality would continue, and a tree falling would make noise, even if there was nobody nearby to hear it. On the other hand, if there were no LIFE entities (including people), not only would the falling tree not make a noise, there would be no tree to fall in the first place. LIFE (*essentia*) creates STUFF, and if there's no LIFE, no STUFF would ever get created. It's a little bit like asking, "If there was never any life on Earth, wouldn't all the buildings eventually fall down?" That question makes no sense because if there was never any life on Earth, there wouldn't be any buildings to fall. LIFE is what creates the buildings in that example, and LIFE is what creates the ongoing reality of the physical universe.

- **Do we have free will?** It sure looks that way. It's almost certain that we participate in creation. As creators, we must have at least some free will. We're participants in the grand scheme of things, not just accidents, spectators, or game pieces.

- **Why is morality important?** Morality, done right, should never be treated as just a collection of old and arbitrary rules designed to do little more than dent your pursuit of pleasure. A workable morality is an answer to the question, "How should I live if I want to be happy?" We, as spiritual beings, have certain innate purposes, and if we don't make some definite progress toward fulfilling them, our lives will suck. A workable morality is a set of guidelines on what to do and how to do it so that our lives don't suck.

- **Is the universe deterministic?** Yes and no; it depends on what you mean. Ultimately, we participate in the creation of our moment-to-moment existence. However, it's possible to create something that then runs on its own for a while. Kind of like a wind-up clock. Once time exists and the clock is created and wound up, it can run in an apparently deterministic fashion for a while, maybe a long while.

Those are just a few of the questions that have been kicked around for years and that *essentia* theory actually comes up with a decent answer for. In addition, there are more. Some of these questions will be taken up within the next chapter or two, or the next book or two (or three). For instance:

- Why is there a universe?
- What is the meaning of life?
- What is the purpose of LIFE?
- What is the difference between right and wrong?
- How should we live?
- Is happiness possible?
- Is there a God?
- If there is a God and God is good, why are we allowed to suffer?
- How do I attain health?
- How should we set up and handle our governments and institutions?
- How do we attain sustainable prosperity for all?

No other model even comes close to answering this number of questions. Not only that, you can use these answers to predict things, things that come to pass. Does it answer every possible question? Of course not. Due to the nature of reality itself, no model of it will ever be able to address everything. There are some things that each person simply has to find out for themselves.

Having said that, I think *essentia* theory is by far the most complete and viable theory currently in existence to explain the vagaries of the subatomic and the human realms. Yet these answers are just a jumping-off point for more things to come.

Now we get to the interesting part.

CHAPTER 20

THE MEANING OF LIFE

I once thought that if I could ask God one question, I would ask how the universe began, because once I knew that, all the rest is simply equations. But as I got older I became less concerned with how the universe began. Rather, I would want to know why he started the universe. For once I knew that answer, then I would know the purpose of my own life.
—Albert Einstein

Water the flowers that grow at your heel.
—Steve Winwood, Jim Capaldi, Chris Wood

Old men have been sitting on mountaintops for a long time, looking for answers and wisdom. Part of what they were looking for is something we now have. We have a new and workable way to view and handle reality. While this is certainly interesting in and of itself, ultimately we need to put it to work bettering conditions for it to have any real meaning, or to bring us wisdom. So, we should work on the question, what is the meaning of life?

Polar Qualities

All activity and existence in this universe, all the actions of living, are the product of the interaction of two opposed (or polar) principles or qualities. The most fundamental of these is the division between "that which creates" (including us) and "that which is created" (the stuff of this universe). "That which creates" and "that which is created" are the *basic* qualities of LIFE (*essentia*) and STUFF (*materia*). The qualities of LIFE are the opposite of the qualities of STUFF because the quality of creating is the opposite of the quality of being created.

To be capable of creation, LIFE must possess certain qualities. The qualities of STUFF follow as a necessary condition of what it is and how it's created. We can validate this by observing how LIFE actually works in this universe.

This fundamental division of qualities provides the basis for all other polar qualities, such as alive vs. not alive (dead); order vs. disorder; awareness vs. unawareness; enthusiasm vs. apathy, as well as many others. All of these divisions of qualities align with the fundamental division of "that which creates" vs. "that which is created." As an example, LIFE and being alive are aligned with "that which creates," while deadness or being dead is aligned with "that which is created." Truth aligns with LIFE, while falsity aligns with STUFF. Order is a quality of LIFE, while entropy or disorder is a quality of STUFF. Doing the right thing is a quality of LIFE. Doing the easiest thing (or nothing at all) aligns with STUFF.

None of these polar qualities exist in pure form in this universe. Nothing in this universe is ever totally true or totally false—there's always a nugget of truth in even the biggest lie. The extremes of these polar qualities combine, and that combination is what forms any quality found in this universe.

It's the combination and interplay of order and disorder that creates any working system. Any system has got some order in it, along with some disorder. If it's working, it has probably got more order than disorder. The way the sequence of events runs in living, disorder exists, then order is put into it to create some sort of working system, which then gradually runs down back toward disorder unless it is stabilized by injections of order. For example, a car gets built up from raw materials (order from disorder), then you buy it, and it immediately starts to run down or decay unless you take positive action to maintain it.

Any quality that exists in this universe is located somewhere between the polar extremes. For instance, while complete order and perfect chaos cannot exist in this universe, all possible combinations of them do, ranging from extremely orderly, to orderly, to some order, to a little order, to not particularly orderly, to a little disorderly, to fairly disorderly, to extremely disorderly, along with all possible steps and gradations in between.

Here's an example of how a few of these polar qualities align:

That which creates	That which is created
Alive	Dead
Order	Disorder
Aesthetic	Randomness
Awareness	Unawareness
Enthusiasm	Apathy
Help	Helplessness
Productive	Useless
Truth	Falsity
Solutions	Problems

As a culture, we've tried materialism. While it does extend our control over STUFF to some degree, it eventually produces failed or failing lives and cultures and is doomed to produce nothing but failed lives and cultures.

Why? Ultimately, it doesn't work. It's an incomplete and unworkable idea.

Why? It doesn't work because it recognizes only STUFF, while in actuality, the values and priorities of LIFE are senior to the values and priorities of STUFF. LIFE is the source of and creates STUFF. Any philosophy that recognizes and works with nothing but the products and byproducts of creation and fails to recognize the creative force itself is bound to end up in the weeds.[62]

Living is a rough-and-tumble activity, and while people may start off true to the values of LIFE, they tend to gradually succumb to the "slings and arrows of outrageous fortune," to the unending grind of STUFF. Going from high to low on the scale seems to be the trajectory of life, although some get to the bottom more quickly than you might think.

Individual manifestations of LIFE (e.g., people) can exist anywhere along a scale from mostly LIFE, to some LIFE, to only a little LIFE and mostly STUFF. If LIFE is operating high on the scale, there will be a tendency to create beautiful, orderly, helpful, or productive STUFF intelligently and with good enthusiasm. People high on the scale will tend to think rationally and seek out truth. You might see high-level artists, entrepreneurs, or craftsmen operating like this.

Lower on the scale, the tendency would be for LIFE to create some STUFF, but it won't be especially beautiful, orderly, helpful, or useful, and they'll probably do it without much enthusiasm. These people might not partake broadly of rationality and are not so much interested in truth as they are in expediency: what gets me through the day or through today's crisis?

Really low on the scale, LIFE might be able to create some small amount of STUFF, but that STUFF will tend to be dead or dying, disorderly, have little or no redeeming aesthetic, be destructive or useless, and they will

---

62. An idiomatic expression that means immersed or entangled in details or complexities. See citation in Notes.

generally do this stupidly or in an indifferent manner. People at this level would tend to think and act irrationally, have trouble recognizing truth if they saw it, and might actually be more comfortable with things that aren't true.

Of course, individuals are individual, and they might do well in certain areas and not so well in others. However, as a general rule, as a person goes downscale and agrees more and more with STUFF and less with LIFE, the STUFF surrounding them gets deader, more disorderly, and more destructive (or destructed), and they start to take on the qualities of STUFF themselves. You can see this. Some people bring a sort of light to a room. They think productively, seek to live with truth, and are boundlessly active. Others seem to live in a perpetual shadow, spend much of their lives in meaningless or destructive activities, and wouldn't be much interested in truth even if it was presented clearly and convincingly. They tend to be stimulus-response instead of aware, operating blindly on sensations without much rationality, and possess significant inertia.[63]

The action of the physical (STUFF), or those who have taken on too many characteristics of the physical, is to wear LIFE (you) down and to make you agree with the physical, one way or another. While this may happen quickly or slowly, it usually appears inevitable. The nature of this universe is to tend toward disorder and destruction, and it can sometimes seem like there's no feasible way to resist. It seems that all roads, after whatever detours might arise, lead inexorably[64] downhill. This is what life looks like to many—all downhill from here.

Even so, there is a way; but you have to get off the downward treadmill. Devoting yourself to wealth (especially unearned wealth), being sexed-up,

63. Inertness, especially with regard to effort, motion, action, and the like; inactivity; sluggishness. See Notes for citation.

64. In a way that is unyielding, unchangeable, or unavoidable. See Notes for citation.

or spending all your time chasing highs, short-term pleasures, and kicks are some of the goals of STUFF. There are lots of others. They're all attractive in some way but ultimately bring sickness, misery, and degradation. *At best,* you end up as nothing more than a custodian of STUFF, conscious of nothing but STUFF, trapped by and becoming STUFF. That's not what you want. You want to be, instead, a free spirit who uses STUFF to accomplish your aims.

You are a spirit and can achieve dominion over the physical. If your purpose is good, by simply continuing to look, learn, and create, no matter the opposition, you can start to create the world you want. It may involve time, dedication, and effort, but it's doable.

The Nature of Reality

Reality is created, to some degree, moment to moment, by all of us. It's our looking, our observations, our interactions, that trigger the change from potential to actual and have some input upon what that actuality might be. While solid and separate from us, reality is not totally immune to our touch.

It's a bit like being in a football stadium at game time. There are lots of people around, they're all contributing their voices to the roar, and that roar and that reality might seem overwhelming and sweep you along. However, no matter how loud or overwhelming, that roar is made up of individual voices just like yours. Your voice is part of what creates the reality inside that stadium. And while you can, you don't have to go along with whatever everybody else says and does.

In some ways, reality is like politics, in that some of us have bigger voices than others. It's also similar in that some of us may advocate for wonderful things, while others advocate for things that might not be so wonderful.

In any case, any one of us can mold or shape reality to be more the way we believe it should be. We can bend it toward more LIFE (or less).

Ever notice how some people have all the luck? How they get the lucky breaks, while others don't? They're bending the universe just that little bit toward them. This is something we can all do if we think and act in accordance with LIFE. We don't have to wait for the universe to serve up a break; we can encourage it to dispense one.

You can be more than just another faceless face in the crowd. You can make your own happiness. In doing so, you might also find that the area around you is safer and saner. Maybe the stadium hooligans don't build up the critical mass they need to go beat people up. Maybe reality changes just that little bit for the better. After all, a crowd, while seemingly all-powerful, is nothing more than a collection of individuals, and if you can change your mind, they can change theirs.

The Purpose of Life

The purpose of life is more LIFE. More LIFE does not necessarily mean more living beings. Having a better life is included in the definition of more LIFE, because a better life involves more beauty, more order, more happiness, more intelligence, more truth, more enthusiasm, and other similar qualities.

The purpose of the physical is to provide obstacles to greater LIFE. It, and those forces and entities aligned with it, will attempt to stop or hinder any attempts to create more life. You could say that, as you stand looking at it, a better life exists on the other side of disorder. Disorder is here—a better life is over there, just beyond it. You'll almost always have to go through or overcome some opposition or distraction to get where you want to go, and in doing so, will gain knowledge and experience.

The bigger the barriers to overcome, the sweeter the victory. And the more life created by your victory, the more meaningful it is.

The way things get better is if we make them better. If we don't actively make them better, they get worse. Disorder is the keynote of this universe.

Purposes and actions should be evaluated in terms of how much LIFE results from them. The acquisition of STUFF, power, and the pursuit of admiration for their own sake are examples of ill-advised purposes because they, in and of themselves, do not generally result in more LIFE. STUFF, power, and admiration are great if they're used as part of a purpose to improve things, but end badly if they're treated as ends in their own right. In fact, the single-minded pursuit of power, money, or admiration almost always results in less LIFE.

What about dictators, psychopaths, and ruthless CEOs, you may ask? Don't they acquire power, money, and admiration? Yes, but while decent people earn their power, money, and admiration, the mad just steal them for a while and then mess it all up in the process. What group, company, or country has ever improved in the long term under ruthless or psychopathic leadership? They may dominate or look dominant for a while but always lose horribly in the end. True happiness is not the work of a moment or a day; it's the work of a lifetime.

Examples of it may be rare these days, but knowledgeable execution of a true purpose will result in plenty of STUFF, power, and admiration. The best way to get these benefits is to be worthy of them. STUFF, power, and admiration tend to flow toward LIFE and a life well lived.

Real education is essential to this process. Otherwise, your good intentions may come down to precious little in the way of actual accomplishments. Real, basic education, which everyone should have, means functional literacy (more about that in the appendix), numeracy,[65] an awareness of cultural history and the true nature of reality, the ability to think clearly and independently, together with any specialties (such as the data of a trade or profession) that might be added to these basics. To properly orient your life, real education is needed so that you can start to break from the crowd, think for yourself, and successfully evaluate your actions and purposes.

The Meaning of Life

So, what is the meaning of life?

The ultimate meaning of life is to overcome obstacles in order to create more LIFE. More LIFE does not necessarily mean more living beings. It could, but more importantly, it means individual beings who are more alive, who partake of more of the qualities of LIFE, and who are thus happier, solve more problems faster, are more productive, more stable, and more aesthetic—more like LIFE itself. The more LIFE that results, the better. The bigger the obstacles to overcome, the better. The most meaning will be found in overcoming the biggest obstacles you can manage, to create as much LIFE as you possibly can.

More LIFE and a better life: That is the road to meaning and to happiness. That is the road on which you learn how to get control of your own life, attain stability, and live it on your own terms. That is the road on which you discover what those terms should be. That is the road to the real you.

---

65. The ability to use numbers, especially in arithmetical operations. See Notes for citation.

CHAPTER 21

WHAT ABOUT RELIGION?

There probably is a God.
Many things are easier to explain if there is than if there isn't.
—John von Neumann[66]

The first gulp from the glass of natural sciences will turn you into an atheist, but at the bottom of the glass God is waiting for you.
—Werner Heisenberg

There can never be any real opposition between religion and science; for the one is the complement of the other.
—Max Planck

Science without religion is lame, religion without science is blind.
—Albert Einstein

Decay is inherent in all component things!
Work out your salvation with diligence.
—Gautama Siddhartha (the Buddha)

---

66. Hungarian and American mathematician, physicist, computer scientist and engineer, 1903–1957. See Notes for citation.

As a race, this is not our first attempt to climb the mountain. There have been other times and places when we made progress, or it looked like we were getting close to the top. However, now, thanks to the advances of science, this is the first time we have a chance to actually approach the summit. Wisdom is within our grasp, if only we reach for it.

Now that we know there can and should be a meaning to life, we should have a look at the tools that are there to help find it and get it happening. Does, or should, religion play any role?

Each one of us is a spiritual being who participates in the creation of the reality about us. While handling the stuff of the material universe is interesting and productive, the ultimate truths and realities of our existence spring from us, as spirits. The most important things we can do are to address those truths and realities, and that is done through religion.

I know that the concept of religion is a hot button for many; and while there are a lot of people who cherish religion, it is seen by others as a sham, or "unproven." However, any address to the spirit and its realities is religious in nature.

In about 1960, Kurt Gödel, one of the most profound thinkers of the twentieth century and the creator of the incompleteness theorem discussed earlier, wrote down a list of fourteen philosophical points he had arrived at after a lifetime of science and math, thinking and logic. Point #14 was, "Religions are, for the most part, bad—but religion is not." Why would he write that?

Maybe some existing religions don't speak to us as they used to; maybe they have lost track of what made them vital and vibrant. Maybe they have come to rely excessively on ritual, interpretation, and dogma. Maybe they've even become corrupted. Maybe so. However, this does not mean that they, at their core, are worthless. It does not discount the central value of spiritual practice in life and living.

So, what is religion? Is it meaningless mythology and dog-eared dogma? No! Religion is supposed to be about us, who we truly are, how life actually works, and how to make something worthwhile out of it. It's supposed to be about truth and using that truth to make a better life for all. Ultimately, it's not about authority, rituals, process, or mysteries. A true religion should be about you and me and how we can make a world. It's the study of us and that which is eternal. It's a study of how life actually works. That's what the holy books have always been trying to teach us.

You are not an animal who thinks; you are a spirit who has a mind and operates a body. Granted, your body is an animal body, but that doesn't make *you* an animal. You should be running the show, not it. This point of view may be a change from the way you have been accustomed to thinking. Looking at living from the viewpoint of a spirit, each of us needs to figure out what that really means, who we really are, how things really work, what our relationships to others and to the world at large should be like, and how to improve all those things. Religion is (or should be) all about how this is done.

Any religion that enhances the LIFE of its adherents—their happiness, fulfillment, and productivity—is valid to that degree, and only to that degree. That religion may be new or ancient; may have a God, many gods, or might not address the issue of God. It may prescribe a rigid morality or something more flexible. It may be deeply into dogma and ritual or may be more concentrated on how and what you're doing here and now.

The important point is not how any one religion is constituted or what it looks or sounds like. What's important is that it works to increase the LIFE of its followers.

The visible part of a workable religion might seem a bit odd or different when viewed from the viewpoint of day-to-day life because . . . it is. Sometimes prophets and wise men seem to speak in nothing but riddles. That's because religion has to do with the realities of LIFE, a creative force that is not of this universe, but that works within it. Day-to-day living is usually concerned almost exclusively with the STUFF of this particular universe. "That which is created" (the physical and those things aligned to it) can never really comprehend or incorporate "that which creates" (the spiritual), any more than horses understand what the rancher is up to. Spiritual realities will usually seem a bit different, out of touch or unrealistic, to those who spend most or all of their time working with, or concerned with, the physical. However, the realities of "that which creates" are senior to the actualities of "that which is created." LIFE transcends this universe and its STUFF.

Religions have tended to be communicated as stories or parables or allegories. Trying to talk to a person immersed in physical reality, with a language that is oriented toward that physical reality, about spiritual realities is like trying to use the language of steam engines to describe electricity. You end up attempting to tell people who understand only fire, heat, and pressure about these magical-sounding, invisible, and intangible force fields, which attract or repel things, and how one can use these magical force fields to make a motor spin. First time through, it probably sounds like BS. However, for anyone who can figure out what's really being said, it opens a door to a whole new world.

Morality

The idea that one should be moral or subscribe to some morality has become *déclassé*[67] in recent times. Some think of it as being pointless or

67. Reduced to or having low or lower status. See Notes for citation.

uncool and reject it. Others try to work out some sort of *ad hoc*[68] minute-to-minute morality, without the benefit of any core or organizing principles. Neither is a good idea. The first is immoral or amoral, the other a pretense. Morality needs to be based on some sort of fundamental truth, not the style *du jour.*[69] Having and applying a workable moral code based on fundamental truths is essential to happiness and LIFE.

Morality is one of the tools of any spiritual practice because it provides some guidelines about how to live a life that's meaningful to you, helpful to others, and forms the basis on which your happiness and a whole civilization can be built. Done right, it's a workable answer to the questions, "How should I live?" or "How should I act?" or "What should I do?" Done poorly, it's an artless[70] substitute for knowing how to live your life.

You *could* have a morality that addresses physical survival only; but if it fails to address the reasons for living and the higher realities of LIFE, such as love, purpose, happiness, aesthetics, joy, and wisdom, it misses the point of why people should survive at all. Most current versions of "morality" have this problem. A code of conduct based solely on physical survival also misses the point that a morality based on these higher realities of LIFE is a necessity in order to achieve any state of living higher than dog-eat-dog physical survival. There is no happiness or progress that is not based on a code of conduct that aligns to some degree or another with the purposes of LIFE. A morality based on false principles, expediency, short-

---

68. For the special purpose or end at hand; also, by extension, improvised or impromptu. See Notes for citation.

69. Fashionable; current. See Notes for citation.

70. Poorly made; inartistic; clumsy; crude. See Notes for citation.

term pleasures, greed, or power hunger leads only to unhappiness and failed societies.

If you don't believe me, take a look around. Do we live in a happy society? Are we happy as individuals? We've obviously got the whole morality thing under control, right? That's why one of the attitudes most prevalent these days is "my life sucks." That's why suicide rates are booming. That's what you get when you can't come up with a workable answer to the question, "How should I live?"

This doesn't need to be the case. It's just that we don't know how to live.

Morality should never be just a bunch of musty old arbitrary practices. It should not be an attempt to bar any and all pleasures. It should not prevent expansion or prosperity. Instead, it should be a set of workable guideposts, which lead you to happiness, fulfillment, and a society in which you're happy to live. Morality is valuable exactly to the degree it accomplishes that, and no more.

The Supreme

In regard to religion, an important question arises. Is there a God?

Ultimately, that's for each individual to find out for themselves.

Having said that, here's an opinion. There are many forms that LIFE takes, from germ to dog to human. All have and express the eternal spark. What distinguishes them is the "amount" of that spark, the amount of vitality, the amount of thinking and responsibility going on. Dogs don't understand the realities of humans—they understand the realities of dogs. Knowing this, who is to say that humans are the highest possible form of spiritual life? In all likelihood, if such a higher form exists, we wouldn't see or comprehend it even if it was right in front of our faces, any more than dogs understand algebra.

Look around. There seems to be an overriding purpose and a guiding hand to this world. I suspect there may be a higher form of LIFE, just waiting for us to grow up and graduate. Maybe there is a God. It sure would explain lots of things. In any case, the journey to discover the truth is, for each of us, the grandest of adventures.

Looking Forward

Each one of us is a spiritual being who participates in the creation of the reality about us. For better or worse, we contribute some part, however large or small, to the creation of the world we see. LIFE transcends STUFF.

A fundamental purpose of LIFE is to make that world a better place, and the tools of religion, including morality, are essential to accomplishing that purpose. Some established religions may have become less vital than they were, but that does not invalidate their core truth. Religion and morality are there to show us the way forward and are valuable to that degree. The way forward may take us much further than we think.

I mentioned a bit earlier a list of fourteen points laid out by Kurt Gödel as the basis of his philosophy. There are a few more points from that list that are of use and interest to us going forward:

- The world is rational.
- Human reason can, in principle, be developed more highly (through certain techniques).
- There are systematic methods for the solution of all problems (also art, etc.).
- There are other worlds and rational beings of a different and higher kind.

- Reason in mankind will be developed in every direction.
- Materialism is false.
- There is a scientific (exact) philosophy and theology, which deals with concepts of the highest abstractness; and this is also most highly fruitful for science.

We have before us the opportunity to turn an evolutionary corner. We have new principles, which we can apply to all the activities of humanity. What other fields could we conquer? What other citadels of mystery might fall to our swords?

CHAPTER 22

CONCLUSION

I believe we are here to do good. It is the responsibility of every human being to aspire to do something worthwhile, to make the world a better place than the one we found.

—Albert Einstein

Be wise and learn my secrets, how disease is healed, how man and beast and plant may talk together and learn one another's missions. Go and live with the trees and birds and beasts and fish, and learn to honor them as your own brothers.

—-message of the last of the Stone Giants[71] to the Haudenosaunee[72]

There comes a point in your life when you need to stop reading other people's books and write your own.

—Albert Einstein

I was one of those annoying kids who always wanted to know why. No matter what the answer was to whatever question I was asking, I always had another "Why?" loaded up and ready to roll.

---

71. A legendary, powerful, hardened-skin people who roamed the wilderness terrorizing others. See Notes for citation.

72. The Iroquois (Native Americans). See Notes for citation.

Years of study and work got me to the point where I think I have some genuinely good answers to questions about life and this world. Frankly, I was surprised to find those answers, and even more surprised when I saw the doors they opened. However, I wasn't so stunned that I failed to barge through those open doors. One doesn't often get a chance like that.

It could seem, at first glance, that questions such as this are just some airy-fairy speculations, which have no particular relation to more important issues like "How do I keep a roof over my head?" or "How do I make more money?" or "How do I get people to respect me?"—all the way to "How do I make it so that my life doesn't suck?" "How do I stop the pain?" "How do I get a life I think is worth living?"

But when we're dealing with these questions, we're dealing with the very basics that underlie this universe and all life. Getting answers to these questions gives us not only the key to handling STUFF but the key to our lives. Once I realized what all this meant and put it to work, my life started to change. Gradually, I got much better at handling physical things. My STUFF didn't break or degrade anymore; it worked; it was good to look at, and I was able to get as much of anything as was needed. Money was no longer an issue; everything just kind of fell into place. It doesn't mean I didn't have to work or think or work things out. What it does mean is it got so that when I tried to work things out, they worked out! And if they didn't, it usually didn't take long to spot what the problem was and get it sorted out.

Also, things in my personal life went just as well. I was able to figure out what I should be doing and was happy to do those things. I was able to get my life under my control, not the control of some government or employer or social media group. My relationships smoothed out, my family life got to be wonderful, and I came to know happiness and experience fulfillment in life.

I became not only rich in STUFF, but rich in meaning and happiness. This was remarkable. Even though my original purpose was merely to satisfy my own curiosity, I realized that I had my hands on something of great value. The issue then became, "How do I workably communicate these ideas?" This data has its roots in advanced science, math, and philosophy, and it's not like most people are going to jump at the chance to learn about science, math, and philosophy just to get a better handle on life. Nor is it going to be easy for them to throw out many of their established (and comfortable) ideas in favor of a whole new viewpoint about life and reality.

How could I talk about this stuff in such a way that it was approachable, made sense, and wasn't horrible to read, without losing the core and the basics of the subject? This book is part of the answer to that question.

The Road to Discovery

Getting all this new data on board has been quite a road. We've been through philosophy; we've found out that we do actually exist, that there can be answers, that workable truth does exist, and that we can prove these things well enough to use them as the basis for our new theory. We've been through science, discovering some of the laws of how our world of STUFF works, and then finding out about how these laws are broken to bits at a subatomic level. We've found out that the information about what happens at a subatomic level, while accurate and usable, has totally shattered all the models of reality we've grown up with, leaving us with mysteries and unanswered questions. We've taken a look at some of the radical new conclusions coming from math and science, which show us that this universe is fundamentally chaotic and partially unknowable, and that its origins lie somehow beyond its boundaries.

Taking all this information and applying tools of logic, such as the principles of simplicity, causal adequacy, and workability, we have undertaken a new explanation, a new model of reality—one which, while perhaps new and unfamiliar in some ways, works better and explains more than all the others that have come before.

Under this model, the actuality of life is that we each play a part in its creation. The ultimate meaning of life is to overcome obstacles in order to create more LIFE. More LIFE does not necessarily mean more living beings. It could, but more importantly it means individual beings who are more alive, who partake more of the qualities of LIFE, and who are thus happier, more productive, more aesthetic, more like LIFE itself.

These data in hand, questions that have troubled humanity since we first began to think might finally have some workable answers, and these answers will alter how we look at life and the world.

LIFE is senior to STUFF. Each of us is a spirit who, in some way, participates in the creation of the stuff of reality. Knowing these things, we can work out a real and workable relationship to STUFF, each other, and our lives. We can come to a better and more fruitful understanding of how to live and what to live for.

This theory represents an important step forward and opens some amazing new doors for all of humanity. However, the real significance of this theory goes beyond a new way of looking at reality; it goes beyond enabling each of us to recognize truth and attain real meaning for our lives. The real significance is much bigger than that.

The real significance of this material is that it puts answers to almost any important question related to living and this universe within our grasp. Knowing the basic data of life and this universe means we can work out the basic data of subjects that have long been unresolved problems for humanity, and use that information to develop what were problems into valuable and workable pursuits. Among the biggest issues we have these days are economics, health, and civics (governments, morals, political systems, etc.). These areas have always been troublesome, but we now have wonderful new tools with which to improve them, showing the way to a higher condition and culture for humanity. We can come up with real axioms for these areas and, with them, create workable sciences out of practices that had previously been shot through with authority, dogma, and opinion.

Why should you care about these things? You should care because they're necessary in order to make all lives (including yours) better. While there are certainly other important bodies of knowledge, I believe these three are the most fundamental in the relationships of humanity to life, to others, and to the physical. You probably have questions about each of them, big questions. And big questions require big answers. Even though each of these topics could itself fill a book, I wanted to at least provide some idea of where we can go. The books covering these topics will come later.

How does knowing basic principles about people and their relationship to the universe change our approach to economics? What could the life of the spirit (or whatever) possibly have to do with economics? What's the relationship?

The most important thing about any pursuit is knowing why you're pursuing it. If you ask people why they're pursuing economics, you get answers like, "I don't know," or "To pay the rent," or "The person who dies with the most toys wins." These are answers, but frankly, they're mediocre. These answers (and others like them) when applied to economics have consistently led to unhappiness, exploitation of the many by the few, and poisoning the environment. Trouble is, these are examples of the best we've been able to do so far.

What we need is a good answer, a real answer, to the question, "What is the purpose of economics?" Answer that one and you open the door for a better economics. Fail to answer it (or fail to ask it), and you end up with a variety of half-workable solutions like we have now. Answering it, you can make economics a field that can result in prosperity, dignity, and meaning for the many instead of just luxury for the few, as well as safety and long life for the planet, instead of treating it like something to be consumed and thrown away.

We believe we have an answer to that question. We know the true purpose of economics. We have used that to establish the rules for a new and improved economics in which all can have a place. This new economics doesn't mean that some won't be or get rich; in fact, we hope some do. But it does mean that, if they follow the true purpose of economics, they won't be doing it by exploiting or excluding the many. We want a rising tide that lifts all boats.

Prior theories of economics were all attempts at attaining this goal, but all failed, some more catastrophically than others. We can take whatever wisdom was inherent in these earlier theories, build on them, and come up with a system that can actually work for everyone.

OK, so maybe there is some relationship between your model of reality and economics. How does this theory relate to health? According to your thinking, the spirit and the body are two different things. Why would one influence the other?

I'll try to answer these questions in order. The relationship between the model of reality and health is similar to the situation with economics. If you don't successfully answer the question "Why do this?" as a matter of first importance, all your subsequent thinking is going to be based on incorrect presumptions, and your conclusions are going to be corrupted as a result. While these conclusions will always contain a certain amount of rightness, ultimately that field of knowledge is going to fail to get you where you want to go.

We can now answer that "why" question, "What is the purpose of health?" With that answer in place, we can bring our other realizations about LIFE to bear and have a better look at how LIFE operates on a most basic level. Once the basics are sorted out, we can use them to work out ways to enhance health and the experience of life, making it more vital and stronger. We can go beyond just trying to maintain an existing (and declining) level of health and move into enhancing or augmenting our physical well-being. We're not looking for cures for illness; we're looking for an enrichment of vitality.

In regard to the second question, yes, the spirit and the body are different. But in us, you could say they're married. It's well known that unhappiness can lead to a sick or weakened body. And a painful or upset body can drag down the happiest of people. So, there is a definite influence from one to the other.

Fortunately, this also means that coming up with ways of improving the condition of the body can not only expand or preserve health, it can also remove some of the barriers to a bettered spiritual state and a better state of mind. Healthier people are, on the whole, happier people. They're also more able to do what they want to do or need to do in order to create a better life for themselves and others.

A significant contributing cause of depression and unhappiness these days is simply poor health. I don't mean by this that people are necessarily physically ill, although many times they are. What I mean is that their vitality is markedly reduced by having to deal with all sorts of nonoptimal issues. People are struggling against a dead weight of environmental toxicity, incomplete nutrition, drugs, poor sleep, and similar issues. It's like trying to maintain the performance and appearance of a fine automobile when you can't get spare parts, the fuel is corrupted, and there's constant acid rain. Things are bound to go downhill.

Fortunately, the way is open for us to do better. It's not really hard, but it does require change.

What could a model of reality possibly have to do with governments and civilizations? They are what they are, right?

No, that's not true, and thinking this has led us to the soup we're in now.

Civilizations are made up of individual people, their beliefs, and their relationships. What do they believe is "right"? What are they trying to accomplish? Believing that "might makes right" and using beliefs of this sort to accomplish world domination will give you a very different civilization from believing that "physical pleasure is the highest aim of life" and trying to create a society where the elite have the time and resources to pursue ever more sumptuous pleasures.

Just like with economics and health, the first and most important question any civilization has to answer is "Why?" Why do we have a civilization at all? What's the purpose? And just like with economics and health, our past has been filled with lots of answers to that question, all of which have been lacking in some way. The existing answers have tended to lead either to some sort of warlord (or warlords) running things, or a debased society that eventually gets conquered by a warlord.

Government is similar. It's also made up of individual people acting on their established beliefs. But why should we have a government at all? Anarchists think we shouldn't. This is another case where the question of "why" has not been satisfactorily answered.

Past experience to the contrary, governments can be useful. But they have to be established and run for the benefit of the governed, or else they become suppressive. Knowing what they're good for, and having some genuine idea of how to accomplish this good, is essential to any successful government. Furthermore, they have to be staffed by people who know about these goals and agree with them; otherwise they inevitably get taken over by opportunists using them to achieve power and wealth at the expense of the governed, like what you see in the world today.

So, knowing what you're about and what you're trying to accomplish is vital to making a success of any government or civilization. To do this, you have to be able to successfully answer the most basic questions related to existence. Prior attempts to work this out were based on incorrect or incomplete basic data, and all have failed.

We can now answer those questions, and answer them successfully. We can know what we're about, what we're trying to accomplish, and how to best go about it. We can establish a new philosophy of government and a new culture, from which can grow new hope for humanity.

The Future

One of the basic theorems that comes from this material is that the institutions of a society are no better or wiser than the individuals who make them up. That includes corporations and governments. This single fact is a large part of the reason why things have been trending downhill recently. You can't build a successful society or government out of broken or suppressed people.

Fortunately, change can come, and not just change for the sake of change, but change for the better. Such a change would be based on a better understanding of life.

Those better understandings and better ideas now exist. What's needed is for people to see or hear them, come to understand them, and from there realize how to use them to make their lives, and all lives, better.

To that end, books taking on each of these subjects have been written and are being readied for publication. One will take on economics, one will take on health, and one will address government and civilization.

These are all challenges, huge challenges. But we now have the ability to meet them and conquer them. It will take a lot of work and it might take time, but there is a road to follow. No more wandering in the desert.

Ready?

APPENDIX

MORE WORDS ABOUT WORDS

This is important.

Even though it's not really about reality, it still bears discussion. As ordinary as it may seem, this may be one of the more important things we could take up.

There is one single cause that is responsible for more miscommunication, miscomprehension, and miseducation and the resulting unhappiness than anything else. It's mostly unknown, even though it's been right under our noses all along.

Proper and exact definitions of words as they are used are *crucial* to understanding.

Back in the mid-1700s, according to *The Philosophical Dictionary*, French writer and philosopher Voltaire said (my translation), "Define your terms, I tell you, or we will never understand each other."

Why would he say that? Granted, he was involved in some of the hottest topics of his day, but this was a very literate person speaking with other very literate people in a language that wasn't anywhere near as diverse in its sources as modern English. They should have been able to sort out what they were talking about without getting too hung up on definitions, right?

Definitions can be tricky. Here's an example of a communication with a missing definition. Let's say I'm your assistant. One day you tell me, "Go out and get me a bouillabaisse." A what? I have no idea what this means;

so, after heroic efforts and some creative thinking, I come back with a French dog.

Here's another example, with a definition that is merely lousy, instead of missing. Let's say you define the symbol "2" as meaning "a number between 1 and 5." If so, what's X in the equation "2+2=X"? It could be anything between 2 and 10.

Finally, here's an example where two people didn't agree on which definition of a word applied in a particular context. Let's say you tell me, "The room was bright with conversation." You meant that there was an abundance of witty conversation. I thought you meant "the room was brightly lit and there was some conversation too."

Any language has definitions that are fairly explicit and well agreed upon, and they're found in dictionaries. But . . . how many words have you looked up in your life? A hundred? Maybe a thousand?

Considering that the average vocabulary of an English speaker these days is about twenty to thirty thousand words, this means you've probably never looked up most of the words you use. One might then wonder, "If you haven't looked up their meanings, how fully and accurately do you actually understand them?"

Realistically, you kinda do and you kinda don't. For most words, you've picked up the general idea here and there, and that's usually been good enough. Sure, there are a few words you don't understand at all; but with most of them, you have a decent idea of what's going on, and that's usually enough to communicate simple or straightforward things. It doesn't require a deep understanding of a language to say hello, talk about dinner, or find the bathroom.

Randomness

In human activities, the distribution of a quality from high to low usually follows a bell curve, like this, where a few at the bottom are deficient, most have enough, and a few at the top have a lot:

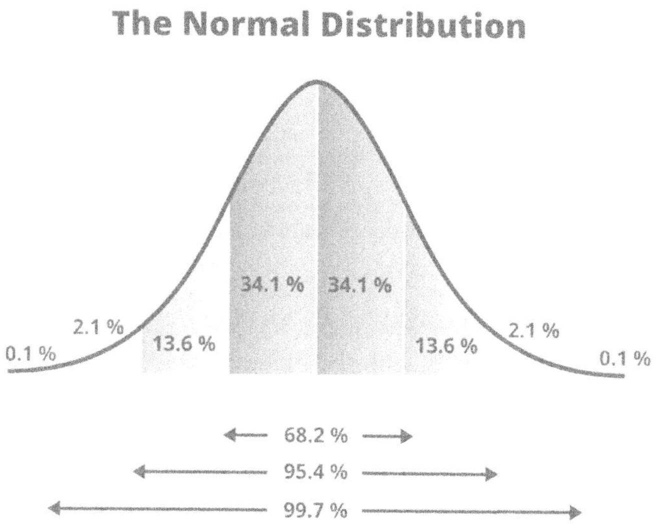

Figure 10: What a normal distribution looks like.
Source: petrroudny – stock.adobe.com.

This curve applies to many of the qualities of humanity and the universe, such as physical traits, cognitive abilities, intelligence, income, performance, and the like. As you can see, there are a few on the high end and a few on the low end, but most are concentrated in the middle. For example, the IQ for the majority of people is between 85 and 115, with only a few higher than that and a few lower.

This curve also applies to your vocabulary and how much of the full meaning of a given word is understood in a working context. It means that some words are better understood than others. The furthest two sections on the left correspond to an incorrect understanding or a small amount of understanding, respectively. The furthest two sections on the far right correspond to excellent, subtle,[73] and total understanding, while the middle sections correspond to adequate understanding.

You may be thinking, "OK, but why are we belaboring this point? I understand most of the words I use fairly well—isn't that good enough?"

Well, let's see. We'll take a fairly straightforward example, and look up the word *account.* You can probably provide one or two different definitions for that word. You might even know that its origin has something to do with the phrase *to count.* However, according to Dictionary.com, that word has no fewer than eighteen separate and distinct definitions, plus a derivation. Granted, some of these may be technical definitions that apply only to a specific field or might be just shadings of a similar meaning, but these shades are shades you could be missing.

Anyway, let's give these six sections in the graph above a numerical value representing a percentage of the total understanding of a given word.[74] *Understanding* for this purpose will relate to "how much of the *full* meaning and flavor of the word did you actually get?" Starting from the left, we'll get (minus) –10 percent, 50 percent, 85 percent, 85 percent, 95 percent, 100 percent. Using these figures, roughly 2.1 percent of your vocabulary will be completely misunderstood (like believing *dawn* means the evening time, or something), while another 2.1 percent will be totally, correctly, and thoroughly understood in all of its implications and shadings of meanings. About 13.6 percent could be a bit muddy or

73. Fine or delicate in meaning or intent. See Notes for citation.

74. I'm basing these percentages on my own observation, and not on any hard data. These are estimates used for the point of illustration.—William Truax.

muddled, while another 13.6 percent will be very well understood, and 68.2 percent will be understood well enough for daily use.

This shows that you're going to understand most words moderately well, a few words very well, and a few more hardly at all. Fortunately, the ones you really misunderstand badly are going to be a small minority of the total.

You might be wondering why I think that this bell curve has any applicability to vocabulary. While many people have noted that a distribution like this is a "normal distribution," why is it "normal"? What makes it "normal," and why would it apply to vocabulary?

This curve is what happens when you leave something to chance. If you have a range of outcomes and those outcomes are chosen with some degree of randomness, this curve is what results.

Even though it may seem odd, I'm not inserting randomness into this conversation randomly. This universe, every particle in it, and every life lived within it partakes of this randomness. Randomness, also known as indeterminacy, is an unavoidable fact of our lives and this universe. A certain amount of randomness is inherent in every single thing we do. It's the way this universe is built. One of the things that it can apply to is language—how much of all the meanings of any given word do you truly understand? By not looking up words, you're bringing a certain amount of randomness into your understanding of the language.

You may say, "Learning a language is not just random. People learn from teachers, from books, and from each other what words mean." Indeed. That's why I've skewed the percentages of understanding given above to the high side. However, while people do learn from each other, that learning process can be incomplete or flawed. As mentioned above, how many words have you actually looked up in your life? Five percent of your vocabulary? Ten percent? Maybe the person you learned from looked up as much as 20 percent, but then how many of those definitions did they teach to you?

Efficiency

Some thoughts require more "communication" to get across than others. Some thoughts are newer, more detailed, more subtle, or more involved than just talking about dinner. These thoughts require a greater depth of understanding of the words involved than straightforward everyday conversation. Social media (as an example) doesn't usually convey a ton of meaning, so simple or fuzzy concepts of words work acceptably well for that. It's only when we get to learning new things, gaining greater depths of understanding, working with a technology, or thinking new kinds of thoughts that we really need a good depth of understanding and the ability to differentiate, which comes with it.

The importance of word definitions varies depending on the significance[75] content of the communication. One could say that a particular idea requires an 80 percent utilization of the full meaning of the words involved, while another only requires 60 percent, and another only 20 percent. New, complicated, precise, or subtle thoughts require more "communication" than simpler thoughts. They require a higher degree of what we could call "efficiency."

So, how does all this affect verbal human communication? Let's say that there's a sentence that conveys a new, important, precise, and/or subtle concept, and that 95 percent of the full and precise meaning of those words in context is needed to completely communicate that concept. Let's also say that sentence contains five words, four of which are 100 percent

---

75. Meaning; import. How much meaning is loaded into that communication; the meaning or implication.—William Truax.

understood by the recipient, and one of which is 80 percent understood. Assuming the originator of the communication fully understood all of the words being used, the total conveyed understanding is 100 percent times 100 percent, times 100 percent, times 100 percent, times 80 percent = 80 percent.[76] Only 80 percent of the total possible understanding got conveyed. And there's no way for us to know which 80 percent was successfully communicated and what got lost or altered.

Hmmmm . . .

What happens if three words in that five-word sentence are 100 percent understood and two are 85 percent understood? You get 72.25 percent understanding. What happens if three words are completely understood, one is 85 percent understood, and one is 50 percent understood? Total conveyed understanding drops to 42.5 percent of total potential understanding. And all that's assuming that one side of the communication (the speaker) understands all the words used well. Bear in mind that the speaker could have some vocabulary gaps as well—what then?

Plot these facts out against enough words and levels of (mis)understanding and you get a curve that looks like this:

76. One hundred percent for each of the first four words, times 80 percent for the final word. That's how the math on this kind of math works.—William Truax.

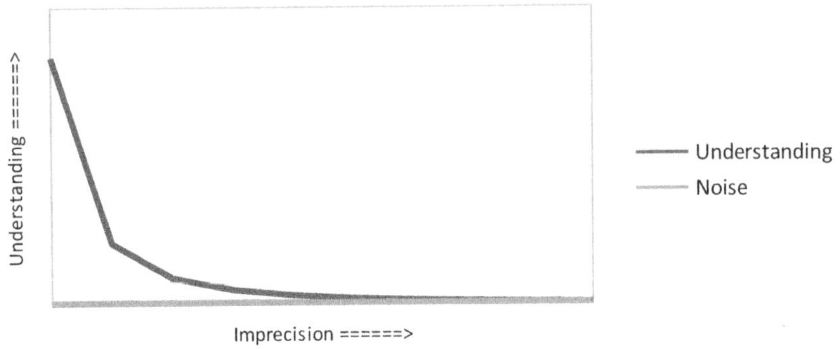

Figure 11: How understanding declines with the imprecision of definitions. Source: William Truax.

The highest point on the upper line represents something like a complete understanding of what was being communicated, while the flat lower line ("noise") represents the level beneath which things "don't make any sense" or don't convey anything new. Notice how fast the upper line gets near the "noise" line? It doesn't take much imprecision to ruin the communication of an important idea. One confused idea from a bad definition can (and will) impair or even ruin the impact of a communication, depending on how far off that one bad definition is.

The example above used a thought that required 95 percent communication efficiency (meaning 95 percent of the meaning of the words involved was needed to successfully communicate the idea). Simpler ideas have a different requirement. If an idea only requires 40 percent efficiency (total possible meaning of the words involved to communicate), e.g., "I'm hungry," then a 40 percent understanding of the words involved could be sufficient to get that idea across. Only if total understanding of the words falls below 40 percent does communication become impaired.

Negative Understanding

There are some other interesting factors that can come into play, like negative understanding. All it takes is one word being grossly misunderstood to turn the total understanding negative, meaning the listener gets something completely different from what the speaker intended. Of course, I suppose you've never seen anybody listen closely to some instructions and then do something completely different. Never.

All it takes is one very bad definition to render a simple communication totally worthless (or worse). And bad definitions are increasingly prevalent in our society as more and more of us are taught, when encountering a new or confusing word, to "figure it out from the context," or maybe just to ignore it altogether and keep on going. While it will probably take more than one word in one sentence to ruin a communication, sometimes the worst can happen. For want of a nail,[77] you know?

Why would I bother to go into this kind of detail? It's not revelatory[78] that badly defined words could impair communication.

What *is* a revelation is finding out just how severe an effect this is, how hard it hits, and what happens afterwards. It only takes a little imprecision or lack of agreement on the meaning of any one particular word to markedly reduce the understanding of an important communication.

---

77. For want of a nail the shoe was lost.
For want of a shoe the horse was lost.
For want of a horse the rider was lost.
For want of a rider the battle was lost.
For want of a battle the kingdom was lost.
And all for the want of a horseshoe nail. See Notes for citation.

78. Having to do with something revealed or disclosed, especially a striking disclosure, as of something not before realized. See Notes for citation.

"Three plus five plus blurteen equals . . . ?" Depending on what you think about "blurteen," you could end up almost 50 percent off the mark.

Even though it may be easier to see the damage bad definitions can do, knowing how it works also highlights just how much upside is available to us. Our ability to communicate important ideas is actually much better than you might believe. In fact any subject, properly understood and communicated, can be communicated successfully. Any subject, if its truths are known and understood, can be learned. However, successful learning depends on proper understanding of the words involved in its communication.

As simple as it may seem, imprecise or incorrect definitions can make human advancement difficult or impossible. A culture is no more nor less than the aggregate understanding of how things work, and incorrect or imprecise words limit or inhibit understanding. Sitting right there all along, plain as day, this effect is one of the primary barriers to civilization. It's one of the primary reasons our current civilization is failing.

Understanding—Lost and Gained

At first glance, verbal communication is pretty simple. You talk, I listen, and I get the idea you're talking about. Or, I write, you read, and you get the idea I'm trying to put across. The whole idea of verbal communication is to convey some sort of an idea from one party to another.

Remember how the word *account* had eighteen different definitions? If I use that word and we're going to communicate with any success, we both have to agree on which of those eighteen definitions is in play this time. So, we need not only precision in our definitions but agreement on which one is getting used in any particular circumstance.

If we have precise *and* agreed-upon definitions for words, we can communicate clearly. If our definitions are vague, missing, or not well agreed upon, our ability to communicate usefully will diminish. For example, let's say we're going to go for a ride in my airplane, and I tell you, "Bring a set of headphones." By *headphones*, I mean wired, noise-canceling headphones that can be plugged into the plane's communication systems. As a pilot, this is obvious to me. But you interpret *headphones* as meaning little wireless earbuds for your cell phone, which is what's obvious to you. But those little wireless earbuds are useless in the noisy cockpit of an airplane. In this situation, the math works out as 100 percent times 100 percent, times 100 percent, times 100 percent, times 0 percent = 0 percent. No comprehension. I might as well have told you nothing at all, for all the good my communication did—it was effectively useless. *Verbal communication is limited by agreement upon the definitions of the words and symbols used, as well as by the precision of those definitions.*

Even more important, the effectiveness of a verbal communication at delivering its intended idea drops dramatically as the precision of, and agreement on, the definitions fall off. All too quickly, the level of understanding of the new incoming communication drops to, or below, the level of things the recipient already knows.

In real life this means there's a huge gap between the potential understanding between people and the actual amount of understanding of important ideas that often occurs. Unfortunately, it only takes a little fuzziness to completely bar the communication of something new or significant. You simply can't communicate 100-watt ideas with 75-watt definitions. They have to be more precise that that. As mis-definitions creep in, the truly significant ideas on which a culture or a technology are based quickly get misunderstood or lost altogether.

Even worse, since much of our own thinking is done with words, what happens if we're kinda fuzzy on the meaning of some of the words we're thinking with? It means that not only will our communication go south, but our thinking will become muddled and unclear. We won't have a decent understanding of what we're thinking about, much less be able to explain it clearly. This muddle obstructs a person's native intelligence. *Clearing up the language and getting words well defined can lead to increases in perception and IQ.*

Verbal communication can (and frequently does) get impaired enough that only simple ideas can be reliably understood and communicated—ideas like "I'm hungry." When this happens, new or complex ideas are either not understood at all, are misunderstood, or are twisted up to fit into the framework of what the listener already understands. Ever seen that happen? This impacts learning, and learning impacts society. A society is going to be no smarter or more capable than the people in it.

You've probably seen cases where some person carefully explained something to another person, only to have the recipient not "get it," or get something *completely* different from what was intended. It's easy to assume this was because the recipient wasn't too bright or wasn't listening, or something. In actuality, it was probably because the speaker and the recipient didn't have enough knowledge, precision, or agreement on what the words in use actually meant. So, the recipient didn't "hear" or didn't "get" what was being taught.

Some words create emotional responses in people, completely apart from whatever understanding they might have of the word itself. These words are "hot-button" words. Words like "nuclear" or "race" are examples of hot-button words. The emotional responses these words churn up tend to steer people in certain directions or toward certain conclusions regardless of how little sense those conclusions make. *Getting words clearly defined decreases the susceptibility of a person to the "hot-button" effect.*

If the meanings of words get vague, people are easily confused and can fall for false data or opinions pretending to be data. They can't tell the difference between bogus data and the real thing. This leads to reactions, emotions, and prejudices taking the place of reason, and ideology taking the place of clear thought. Foolishness can occur on a large scale because people are cut off from the experience and wisdom of those who have gone before. There can also be a decline in overall intelligence because an increasing percentage of the population is cut off from actual education.

All because we didn't think it was important to clearly define our words.

The new idea here is not just that lousy definitions impair communication. We were somewhat aware of that already. What's new is the idea that *the quality of communication decreases, not as the sum, but as the product of the inadequacy of all definitions of both speaker and listener in a given communication.* It's not a nice, linear decrease in understanding—it's where understanding falls off a cliff.

What's new is the recognition of how fast understanding drops and how fast teaching and communication become ineffective in the presence of poor definitions.

What's new is the recognition of how fast lousy definitions can ruin society. It's not a gradual falloff. It's hard and fast. You may have noticed that, as education has become less effective and literacy has fallen off, the pace at which human societies are deteriorating has increased.

Because the falloff in comprehension is so fast, ideas that require a high level of communication can be made "uncommunicate-able" by anything more than a little imprecision in the definitions. If you wanted to limit the advance of a culture or lose some knowledge, muddying or altering definitions would be a good place to start. Bad definitions effectively put a limit on how well we can think and communicate, and are one of the biggest factors in the dumbing down and increasing divisiveness of our societies. It's probably also one of the major reasons why the hard sciences have generally had a better time of it than the soft sciences. They have defined their words and terms more closely and so understand their subject material better and communicate it with better clarity.

Fortunately, knowing this and some of the ideas mentioned earlier in the book, we can open the door to a new age.

- Any subject that exists in this universe can be known.
- The entire and proper meaning of words can be learned.
- Any subject that can be known can be embodied in properly understood words.
- Any subject that can be embodied in words can be communicated in words, which can then be learned and understood by another.

Words clearly understood and precisely used are vital to thinking, communication, education, and culture. Communication is the single most vital aspect of our existence, and the most vital aspect of verbal communication is definitions.

NOTES

Preface

1. Dictionary.com, "paradox," accessed January 19, 2025, https://www.dictionary.com/browse/paradox.

2. Caltech Science Exchange, "What Is Quantum Physics?" accessed January 19, 2025, https://scienceexchange.caltech.edu/topics/quantum-science-explained/quantum-physics.

3. *Merriam-Webster Dictionary*, "legerdemain," accessed January 19, 2025, https://www.merriam-webster.com/dictionary/legerdemain.

Introduction

4. *Encyclopedia Britannica*, "Max Planck," accessed January 19, 2025, https://www.britannica.com/biography/Max-Planck.

5. Dictionary.com, "inflection point," accessed January 19, 2025, https://www.dictionary.com/browse/inflection%20point.

6. *Merriam-Webster Dictionary*, "materialism," accessed January 19, 2025, https://www.merriam-webster.com/dictionary/materialism.

Chapter 1

7. Dictionary.com, "word," accessed January 19, 2025, https://www.dictionary.com/browse/word.

8. Dictionary.com, "product," accessed January 19, 2025, https://www.dictionary.com/browse/product.

Chapter 2

9. *Merriam-Webster Dictionary*, "parable," accessed January 19, 2025, https://www.merriam-webster.com/dictionary/parable.

10. Dictionary.com, "allegory," accessed January 19, 2025, https://www.dictionary.com/browse/allegory.

11. Dictionary.com, "passé," accessed January 19, 2025, https://www.dictionary.com/browse/passé.

12. Dictionary.com, "problematic," accessed January 19, 2025, https://www.dictionary.com/browse/problematic.

13. Dictionary.com, "put paid to," accessed January 19, 2025, https://www.dictionary.com/browse/put%20paid%20to.

14. Science-Atlas, "Singularity," accessed March 9, 2025, https://science-atlas.com/space/exactly-what-is-a-singularity/.

15. Dictionary.com, "inform," accessed January 19, 2025, https://www.dictionary.com/browse/inform.

Chapter 3

16. Werner Heisenberg, *Across the Frontiers*, trans. Peter Heath (New York: Harper & Row, 1974), 132.

17. Dictionary.com, "Sir Isaac Newton," accessed January 19, 2025, https://www.dictionary.com/browse/sirisaacnewton.

18. Author's comment.

Chapter 4

19. Dictionary.com, "thermodynamics," accessed January 19, 2025, https://www.dictionary.com/browse/thermodynamics.

Chapter 5

20. Author's comment.

21. Wikipedia, "Francis Bacon," last modified January 23, 2025, https://en.wikipedia.org/wiki/Francis_Bacon.

22. Dictionary.com, "proposition," accessed January 19, 2025, https://www.dictionary.com/browse/proposition.

Chapter 6

23. Dictionary.com, "philosophers," accessed January 19, 2025, https://www.dictionary.com/browse/philosophers.

Chapter 7

24. *Merriam-Webster Dictionary*, "enzymes," accessed January 19, 2025, https://www.merriam-webster.com/dictionary/enzymes.

Chapter 8

25. Dictionary.com, "empiricism," accessed April 8, 2025, https://www.dictionary.com/browse/empiricism.

26. Wikipedia, "William of Ockham," last modified April 10, 2025, https://en.wikipedia.org/wiki/William_of_Ockham.

27. Wikipedia, "hypothesis," last modified January 23, 2025, https://en.wikipedia.org/wiki/hypothesis.

28. Author's comment. For more information, see Wikipedia, "Geocentric model," last modified January 23, 2025, https://en.wikipedia.org/wiki/Geocentric_model.

Chapter 9

29. Dictionary.com, "brook," accessed January 19, 2025, https://www.dictionary.com/browse/brook.

30. Dictionary.com, "creed," accessed January 19, 2025, https://www.dictionary.com/browse/creed.

Chapter 10

31. *Merriam-Webster Dictionary*, "positivism," accessed January 19, 2025, https://www.merriam-webster.com/dictionary /positivism.

32. Collins English Dictionary, "tautology," accessed January 19, 2025, https://www.collinsdictionary.com/dictionary/english/tautology.

33. Wikipedia, "Stephen Hawking," last modified March 7, 2025, https://en.wikipedia.org/wiki/Stephen_Hawking.

34. Wikipedia, "John Ellis," last modified January 23, 2025, https://en.wikipedia.org/wiki/John_Ellis.

35. Wikipedia, "Roger Penrose," last modified January 23, 2025, https://en.wikipedia.org/wiki/Roger_Penrose.

Chapter 11

36. Dictionary.com, "electromagnetism," accessed January 19, 2025, https://www.dictionary.com/browse/electromagnetism.

Chapter 12

37. Wikipedia, "Thomas Young (scientist)," last modified January 21, 2025, https://en.wikipedia.org/wiki/Thomas_Young_(scientist).

Chapter 13

38. Dictionary.com, "catalyst," accessed January 19, 2025, https://www.dictionary.com/browse/catalyst.

Chapter 15

39. Dictionary.com, "indeterminacy," accessed January 19, 2025, https://www.dictionary.com/browse/indeterminacy.

40. Dictionary.com, "unalienable." Accessed January 19, 2025, https://www.dictionary.com/browse/unalienable.

41. Dictionary.com, "ginormous," accessed January 19, 2025, https://www.dictionary.com/browse/ginormous.

Chapter 16

42. Dictionary.com, "express," accessed January 19, 2025, https://www.dictionary.com/browse/express.

43. Wikipedia, "Church-Turing thesis," last modified November 8, 2024, https://en.wikipedia.org/wiki/Church–Turing_thesis.

44. Dictionary.com, "derive," accessed January 19, 2025, https://www.dictionary.com/browse/derive.

45. Dictionary.com, "confounding," accessed January 19, 2025, https://www.dictionary.com/browse/confounding.

Chapter 17

46. Dictionary.com, "ellipsis," accessed January 19, 2025, https://www.dictionary.com/browse/ellipsis.

47. Dictionary.com, "kludges," accessed January 19, 2025, https://www.dictionary.com/browse/kludges.

Chapter 18

48. Dictionary.com, "inference," accessed January 19, 2025, https://www.dictionary.com/browse/inference.

49. Dictionary.com, "extant," accessed January 19, 2025, https://www.dictionary.com/browse/extant.

50. Author's comment.

51. The Free Dictionary, "transcend," accessed January 19, 2025, https://www.thefreedictionary.com/transcend.

52. Dictionary.com, "postulate," accessed January 19, 2025, https://www.dictionary.com/browse/postulate.

53. Author's comment.

54. Online Etymology Dictionary, "essential," accessed January 19, 2025, https://www.etymonline.com/word/essential.

55. Online Etymology Dictionary, "material," accessed January 19, 2025, https://www.etymonline.com/word/material.

56. Dictionary.com, "polar," accessed January 19, 2025, https://www.dictionary.com/browse/polar.

57. Dictionary.com, "continuum," accessed January 19, 2025, https://www.dictionary.com/browse/continuum.

58. *Merriam-Webster Dictionary*, "Murphy's Law," accessed April 23, 2025, https://www.merriam-webster.com/dictionary /Murphy%27s%20Law.

Chapter 19

59. Author's comment.

60. Dictionary.com, "artifact," accessed January 19, 2025, https://www.dictionary.com/browse/artifact.

61. Dictionary.com, "entropy" accessed January 19, 2025, https://www.dictionary.com/browse/entropy.

Chapter 20

62. Wiktionary, "in the weeds," accessed January 19, 2025, https://en.wiktionary.org/wiki/in_the_weeds.

63. Dictionary.com, "inertia," accessed January 19, 2025, https://www.dictionary.com/browse/inertia.

64. Dictionary.com, "inexorably," accessed January 19, 2025, https://www.dictionary.com/browse/inexorably.

65. Dictionary.com, "numeracy," accessed January 19, 2025, https://www.dictionary.com/browse/numeracy.

Chapter 21

66. Wikipedia, "John von Neumann," last modified April 24, 2025, https://en.wikipedia.org/wiki/John_von_Neumann.

67. Dictionary.com, "déclassé," accessed January 19, 2025, https://www.dictionary.com/browse/déclassé.

68. Dictionary.com, "ad hoc," accessed January 19, 2025, https://www.dictionary.com/browse/ad_hoc.

69. Dictionary.com, "du jour," accessed January 19, 2025, https://www.dictionary.com/browse/du-jour.

70. Dictionary.com, "artless," accessed January 19, 2025, https://www.dictionary.com/browse/artless.

Chapter 22

71. David Cusick, *Sketches of Ancient History of the Six Nations* (Lewiston, NY: Self-Published, 1828), 20.

72. Peter G. Ramsden, "Haudenosaunee (Iroquois)," *The Canadian Encyclopedia*, last modified May 15, 2024, https://www.thecanadianencyclopedia.ca/en/article/iroquois.

Appendix

73. Dictionary.com, "subtle," accessed January 19, 2025, https://www.dictionary.com/browse/subtle.

74. Author's comment.

75. Author's comment.

76. Author's comment.

77. 13th-century proverb, popularized by Benjamin Franklin in *Poor Richard's Almanack* (1758).

78. Dictionary.com, "revelatory," accessed January 19, 2025, https://www.dictionary.com/browse/revelatory.

ABOUT THE AUTHOR

William Truax (pronounced "true-ax") is a true Renaissance man, whose diverse achievements and interests span multiple fields. His insatiable curiosity and drive to understand the world have led him to explore profound questions about life and reality, which he delves into in his book *Reality 2.0*.

As an entrepreneur, tax law expert, athlete, musician, researcher, and writer, Truax has made significant contributions in each of these areas. Remarkably self-educated in the law, he became one of approximately 350 non-attorneys admitted to practice before the US Tax Court since 1945.

A part-owner of a successful solar power business and a financial management company in Los Angeles, California, his entrepreneurial spirit and analytical skills underscore his ability to navigate complex industries. These qualities have also fueled his exploration into some of the unresolved mysteries of science and the nature of reality.

An accomplished amateur athlete, he has participated in multisport races, finishing second in his age group twice in long-distance events at the USA Duathlon National Championships and placing eighth at the World Duathlon Championships in 2018.

William Truax self-funded his education, earning a degree cum laude in music with minors in physics and philosophy. His musical talent has led him to perform with major symphony orchestras, blending scientific curiosity with artistic expression.

Beyond his professional and athletic endeavors, Truax has contributed to the literary world as a researcher for several best-selling books. His ability to synthesize complex information has enriched his own writing and his quest for understanding the universe.

An enrolled member of the Saint Regis Mohawk Tribe, he brings a rich cultural heritage to his multifaceted life and career. Growing up with one foot in Native American culture and the other in European American culture, he possesses unique insights into both spiritual and materialistic perspectives on life, enriching his examination of the profound questions tackled in *Reality 2.0.*

In *Reality 2.0,* William Truax embarks on an exploration of the meaning of life and reality by examining scientific phenomena, particularly atomic physics. His journey from curiosity to deep investigation provides readers with a compelling narrative that bridges science, philosophy, and personal insight, aiming to offer a realistic explanation of the mysteries that have puzzled scientists and laypeople alike.